Saving the CBC

Balancing Profit and Public Service

PRAISE FOR WADE ROWLAND'S *SAVING THE CBC*

This book should be read by everyone who gives a damn about Canada and the publicly owned broadcaster that unites us in telling our own stories on radio and television. Wade Rowland convincingly documents the slow, politically-directed erosion of the CBC, and he has the expertise to show us how to save, and expand, this vital component in Canadian life. Will we listen to him? I hope to God we have enough sense to do so.

— Farley Mowat

Consider this an impassioned polemic — "debate" is far too sedate — ignited by the CBC's degradation in recent years and fed by cold rage against the main culprits, yet with a surprising optimism about future possibilities.

— Rick Salutin, author and columnist at The Toronto Star

This is a thoughtful and timely roadmap to guide Canadians who still love public broadcasting but who despair of the present condition of the CBC. Instead of a lament, we now have a plan that can make our CBC a model for how a public broadcaster can inspire, attract and engage us all. You must read this book: Wade Rowland's vision can restore a CBC we can be proud of again.

— Jeffrey Dvorkin, Director, Journalism Program, University of Toronto (Scarborough)

Wade Rowland understands public service values and knows the CBC well, especially English Television. His book makes an insightful contribution to a necessary public debate about our most important cultural institution, and his recommendations are largely aligned with the priorities of the 175,000 Canadians who support our work.

— Ian Morrison, Spokesperson, Friends of Canadian Broadcasting

If you're looking for the first principles required for effective public broadcasting in Canada in the twenty-first century, Wade Rowland has articulated them here with clarity and eloquence. No excuses left for failure to act — except for that most Canadian of realities: the lack of political will.

— Kealy Wilkinson, Broadcast Consultant and Executive Director, Canadian Broadcast Museum Foundation.

LINDA LEITH
PUBLISHING

Cover design: Debbie Geltner
Cover image: Krister Shalm
Book design: WildElement.ca
Author photo: Christine Collie Rowland

Legal Deposit Library and Archives Canada
et Bibliothèque et Archives nationales du Québec

Printed and bound in Canada by Marquis Book Printing Inc.

LIBRARY AND ARCHIVES CANADA CATALOGUING IN PUBLICATION

Rowland, Wade

Saving the CBC: Balancing Profit and Public Service / Wade Rowland

Includes bibliographical references.
Issued also in electronic formats.
ISBN 978-1-927535-11-0

1. Public broadcasting—Canada. 2. Canadian Broadcasting Corporation. I. Title.
HE8689.9.C3R67 2013 384.540971 C2013-900042-9

Linda Leith Publishing Inc.
Singles essay: Media & Communications
P.O. Box 322, Station Victoria, Westmount Quebec H3Z 2V8 Canada
www.lindaleith.com

Saving the CBC

Balancing Profit and Public Service

WADE ROWLAND

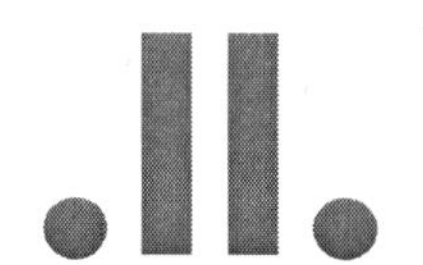

To the spirit of public broadcasting,
and those who struggle daily to
preserve it.

Unlike the Internet, broadcasting is not about supplying a library to which the public has access. Broadcasting assembles a congregation. It is comparable to a concert hall or other meeting halls in our larger cities. Broadcasting is designed to provide a communal experience, an experience that helps build consensus by its very nature, a consensus that should impose the disciplines on talent that ensure that its standards will be high enough to serve that function.

Richard Nielsen, 2012

CONTENTS

Introduction

This is a book about public service broadcasting, and in particular about a crisis facing Canada's public broadcaster, CBC/Radio-Canada. Familiar to generations of Canadians as "the mother corp," the Canadian Broadcasting Corporation has been debilitated by radical bloodletting at the hands of successive federal governments. Between 1985 and 2010, while total federal government expenditures rose by about fifty percent, funding for the CBC was slashed by nearly two-thirds.[1] In the decade of the 1990s alone, the corporation lost a third of its public funding and personnel. Whether this took place in the cause of deficit-reduction, or for ideological reasons, or simply through neglect, is by now a moot point. Today, the corporation faces new fiscal problems that it simply cannot survive. It has become clear that if Canada is to retain a public broadcaster worthy of the name, the CBC will have to be fundamentally reformed, and soon. Previous attempts by corporation managers to adapt the service to shrinking revenues have failed. Current plans to reintroduce advertising on radio would be a disaster. Something different is needed.

The CBC currently operates on a budget of about $1.5 billion a year, roughly two-thirds of which comes from a federal subsidy (its "Parliamentary appropriation"), and the rest from advertising on its television services and website, cable pass-through fees, and other private sources.[2] With that revenue it supplies Canada with two trans-continental television networks (one in each official language);

five cable specialty channels including all-news services in French and English; extensive websites in both languages, including a digital music service with 40 music streams; four continent-spanning terrestrial radio networks in French and English, and channels on the Sirius satellite service. In the north, the CBC broadcasts in six native languages on both radio and television. Until mid-2012 it also maintained a radio service aimed at overseas audiences, Radio-Canada International. All of this involves staffing and maintenance of studio facilities in most major urban centres along with the 1,200 associated transmitters. As of this writing, CBC employs 7,285 full-time staff, 456 temporary full-time staff, and 979 contractees.

To put the $1 billion CBC subsidy in perspective: it is less than eight percent of total revenues for television in Canada and less than twelve percent of total public expenditures on culture (Savage 275).

MOMENT OF TRUTH

There are several possible scenarios for the imminent collapse and possible restructuring of the CBC, all of them involving a drop in already precarious revenues below sustainable levels. The most obvious possibility is a simple decision by the federal cabinet to impose further substantial cuts to the public broadcaster's Parliamentary appropriation, something it could do at any time. But the most likely is this: sometime during the next two years it is probable that the CBC will lose the bidding contest for the rights to broadcast NHL hockey to one of the country's wealthy commercial television networks. When that happens, the corporation will face an unprecedented financial crisis, because *Hockey Night in Canada*, a CBC television mainstay since 1952, provides more than forty percent of total advertising revenue and accounts for roughly 350 hours of prime time broadcasting, a Canadian-content chasm that will have to be filled.

Normally, faced with a crippling financial shortfall of this kind, the obvious response would be to turn to the federal government for emergency assistance. Pleading their case, CBC executives would point to the pitifully low current subsidy, and the historic underfunding of public service media in this country. Both Liberal and Conservative governments have in the past been responsible for the starvation-level financing that afflicts the public broadcaster, but the present Conservative regime in Ottawa is less sympathetic to the idea of public broadcasting than any federal administration in memory. Prominent members of the party, including the Prime Minister, are on record as favouring privatization of the CBC, or at a minimum, some of its functions — Radio 2 being a preferred target, along with English-language over-the-air television service. There is every reason to expect that the government will use the financial crisis as a reason to radically alter the character of the CBC, either through direct intervention in the form of privatization, or a continuation of the kind of benign neglect that has taken the corporation to the brink– opponents of public broadcasting need only stand by and watch it twist in the wind to see their goal realized.

It is the rapid approach of this moment of truth for the CBC/Radio-Canada and the idea of public broadcasting that has provoked this book. It is my contribution to the chorus of voices raising the alarm and insisting on the preservation of what is widely regarded as the cornerstone of a uniquely Canadian culture. I believe the crisis offers a singular opportunity to restructure the nation's media environment in a way that finally acknowledges the fundamental distinction between the goals and purposes of for-profit commercial broadcasting, and the public service model that ought to define the CBC. The new structure would give private broadcasters what they have long lobbied for, and give the Canadian public what it wants, needs, and deserves from its media system. The resources needed to achieve this transformation – human, financial and technical – are already in place. All that is needed is a political strategy.

Full disclosure: I am a man of modest but comfortable means, grown-up family, diverse interests, and eclectic tastes. I like to travel when I can. I read a lot. I spent much of my adult life as a working journalist, first in newspapers and then in network television at both CTV and CBC. I have been a senior manager at both networks. My hair is grey. I have supervised, written, assigned, or otherwise participated in the production of something in excess of two thousand network TV newscasts and sundry other current affairs programs.

I am a life-long, inveterate listener to CBC radio, and a somewhat less loyal and admiring viewer of CBC television. I heartily agree with *The Globe and Mail*'s Michael Valpy who wrote fifteen years ago that: "If we don't have a CBC, we don't have a country.... Without Canadian programming and films, without the broadcasting of Canadian stories, Canadian citizens will not be able to shape Canadian solutions to social and political problems, will not be able to transmit their values to their children."

As a kid raised in the relative isolation of Regina and Winnipeg, it was the CBC that made me a Canadian rather than a Manitoban or whatever residents of Saskatchewan are properly called. I have, without question, learned as much from CBC programming as I did from my undergraduate university education. As a long-distance commuter, I owe my sanity to being able to tune out traffic and tune in to something intelligent to listen to on my car radio as I inch along the 401 or the Don Valley Parkway. When travelling abroad I keep up to date with goings-on back home via CBC's web services, or Sirius.

I am the kind of CBC listener and viewer that corporate management has in recent years decided to jettison in its sad struggle for survival, in its drive to recruit the younger audience that is more attractive to advertisers on television – and politically necessary to arguments for the continued "relevance" of radio. I am the "wrong" demographic, and as such subject to the harsh, lifeboat triage of business logic. I am not at all bitter about this: on the contrary, I can well understand how CBC managers have been driven to take drastic measures, given their desperate financial situation. Who could argue against trying to engage younger viewers and listeners?

What is also understandable, though less forgivable, is the CBC's obsessive fixation on what are known as audience ratings, metrics of success that were invented by and for the commercial media market. This to me reflects a fundamental confusion over the place of public broadcasting in the wider media industry. It can be traced directly back to the petty refusal of successive federal governments to provide stable financing for the corporation, forcing it, year by year, to rely ever more despairingly on advertising revenue, and thus, on ratings.

In my time at CTV, I lived through the early, most ardent years of neo-liberal "rationalization" in media all over North America and Europe, in which managers with business-school backgrounds usurped an earlier generation of leaders steeped in radio, television, and newsprint rather than accounting. In the space of a decade, the notion of overarching social responsibility that had played so prominent a part in the ethos of early commercial radio and television was eradicated. At CTV, the nemesis was a clever young MBA with a track record in cigarette and convenience food sales who was hired to preside over the network and who was openly contemptuous of notions of public service. A technocrat par excellence, he cared only for the bottom line. He demanded that the news division become a profit centre, a more friendly environment for advertisers. Means became ends.

Later, as a manager at CBC television news, I was witness to the crippling effect political venality can have on the morale and performance of even the best, most public-spirited people in a great institution kept perpetually on life-support. Its complexity and fragility made it a manager's nightmare. It seemed to me a construction like Newton's wooden bridge at Cambridge: so cunningly made without need of bolts or nails that when university fellows took it apart to see how it stayed up (the legend says), they couldn't put it back together again.

These days, as a professor of media studies at York University in Toronto, I lecture to theatres full of undergrads and conduct seminars for graduate students, mainly in the ethics of communication, which may sound like an arcane niche but really is central to understanding media, as I hope you'll agree by the end of this book. One of the things I've learned from my second-year students over the past few years is that many of them — perhaps a majority — have no clue what the CBC is or what its purpose is. Public service broadcasting is a foreign concept to them. This has been a scary revelation. Closely related, and also frightening, is the fact that many of them have little understanding of important political, legal, and cultural distinctions between Canada and the US — they know the American constitution and legal system better than their own, and most of their cultural touchstones are American as well. This, it seems clear, is the result of having grown up in a media environment saturated by American programming: Canada's private television networks broadcast virtually wall-to-wall American programming in the prime time territory of 8 p.m. to 11 p.m., Monday to Friday, and have always done so. Commercial radio has little to offer other than pop music and bombast.

While the crisis looming before the CBC threatens its existence, it may also offer an unprecedented opportunity, if it is managed properly. It would be a tragedy if the inevitable debate over the survival of public broadcasting were to take place in the absence of research-based information. It is this body of fact that is essential to countering the industrial-strength, free-market (and often fact-free) rhetoric of advertising and of the private broadcasting industry's lobbyists in Ottawa.

The intent of this book is to lay out some of the information essential to drawing some reasonable conclusions about what ought to happen to the CBC, how it should be governed and financed, and what we as citizens should expect of it in terms of performance. The central theme concerns the question of how we assess the value of media, and in particular a public service medium like the CBC. Do we place our trust in audience ratings, or is there a better way?

The best initial clue I can offer to this conundrum is one contained in an old Zen fable:

A man is racing along a road, clinging to the back of a furiously galloping horse. A friend watching from the side of the road calls out to him, "Where are you going?"

"I don't know!" the rider shouts back. "Ask the horse!"

As with any Zen fable, the meaning depends on the context. For me, it is a warning about the dangers of being captured by your tools, about letting metrics set the agenda, about losing sight of what's important.

This book begins with the question: public broadcasting — what is it? And it moves on to ask how we can gauge its successes and failures, on the assumption that the horse cannot tell us, no matter how earnestly and methodically we interrogate it.

1

What is public broadcasting?

What is public broadcasting? And why should we care about it? The first question is easy to answer. We know exactly what public broadcasting is because it's a social invention, well documented in official archives and in the writings of its creators. It's a question for which there's a factual answer. In a nutshell, public broadcasting is a form of electronic mass communication that is designed to serve its audience as citizens rather than as consumers; it sees its viewers and listeners as a public, a *demos*, rather than a market. Its purpose is to enhance public life, and enrich individual lives, rather than to serve advertisers. Its goals are cultural, rather than economic.

The reasons why we should care about public broadcasting are less immediately obvious. That's because they involve value judgments, and thus step into moral territory. Moral judgments are not impossible to arrive at and to have confidence in, but it can be hard work getting there. Nevertheless, we do it all the time. It may be our defining characteristic as humans. Our public lives are organized around our most reliable and enduring judgments of this kind — about justice and equity and fairness and tolerance and compassion. It's a cumulative process that takes us, for example, from emancipation of slaves, to child labour laws, to universal suffrage, to animal welfare laws, to the abolition of capital punishment, to gay rights, and onward.

There are two ways to look at broadcasting, or any other mass medium (technically, any and all forms of large-scale, one-to-many communication, including movies, newspapers, books, etc.). One is as an industry that through various organizational, technological, and financial means produces a product for public consumption. The other way is to consider the media from a broadly moral perspective: to examine both the media themselves and their content – how it's created and consumed – as revealing expressions of the society that fashions them and which they serve. Both content and the ways in which the media industries are structured and regulated, can reveal much about a society's values and aspirations.

For example, a television or radio program might be made for the purpose of assembling an audience by promising to amuse them, with the further purpose of providing a large number of viewers or listeners for an advertiser's sales pitch. In other words, to make money. Or it might be produced and broadcast in the hope and expectation that it will move, amaze, enlighten, touch, or inspire its audience. Either way, not just the content, but the intent of the enterprise producing it say something about the society in which it functions. Thus, the debate over the fate of public broadcasting speaks to more than simply technological change or audience fragmentation, or concentration of ownership, or any of the other institutional issues that typically distinguish the debate. It says something about who we are.

The notion of public service is deeply embedded in the history of broadcasting for reasons related to the way the technology works. Both radio and television transmit information in the form of radiation within certain portions of the electromagnetic spectrum, a field that includes everything from radio waves through infra-red to visible light up to x-rays and gamma rays. Since the discovery of the field and its properties in the late nineteenth century, the part of the spectrum capable of being used for practical, wireless communication has been thought of as a public resource. Radio waves are said to travel "through the air" or "over the air," and the air belongs to no one and everyone. However, because only a limited part of the electromagnetic field is suitable for radio transmission over long distances, the problem of real estate allocation arose early in the history of radio. If transmitters were to avoid interfering with one another's signals (as they frequently did), they had to operate on different radio frequencies or wavelengths, and the allocation of those frequencies had to be organized both through national regulation and international conventions. But given the public nature of the resource being used ("the airwaves"), users were granted the rights of a lessee, rather than an owner, to the portion of spectrum assigned to their operation. Thus a radio station would be granted a licence to operate on a certain frequency, on the understanding that it was being given privileged access to a public resource. And, from the earliest days of radio, it was understood that commercial radio operations owed the public some form of service in return for this privilege.

In its broadest conception, public service broadcasting (public broadcasting, for short) is an idea derived from the sociologist's definition of societies as "imagined communities." We create and maintain communities both large and small out of a desire to provide the necessary conditions for a civilized life. One of the founders of modern sociology, the estimable Emile Durkheim, saw this kind of social cohesion as being achieved through two contrasting approaches: organic solidarity — bottom-up, consensus-building, democratic, communicative; and mechanical solidarity — top-down, instrumental, paternalistic, authoritarian.

Both strategies, the organic and the mechanical, can be found at work in various combinations in any nation, and the media play important facilitating roles. Radio, for example, was discovered to be both a powerful tool in organizing the Fascist societies of the

1930s and an invaluable aid in mobilizing resistance during World War II. In both cases the medium was primarily an instrument for disseminating propaganda, an essentially authoritarian undertaking in support of mechanical solidarity. But radio, like other media, can be adapted to serve more democratic ends: to be an instrument of organic solidarity and of true communication in the sense of an exchange of ideas and viewpoints. Radio that is managed at arms-length from the state, by professionals, and according to transparent guidelines that allow for plenty of public feedback and participation is no longer an authoritarian instrument, but a facilitator of organic solidarity built on consensus. It provides a public space, a forum, in which the issues of concern to a community can be defined and discussed, with the hope of achieving a consensus opinion and bringing about needed change.

Since the 1950s, television has displaced radio as the most powerful instrument in this democratic undertaking of building a moral community of citizens who share important values. And in our contemporary, media-saturated world, it is safe to go further than community-building to say, as media ethicist Roger Silverstone does, that the media "filter and frame everyday realities... providing touchstones, references, for the conduct of everyday life, *for the production and maintenance of common sense*" (6).

The production and maintenance of common sense is quite obviously an enterprise that carries with it enormous moral and ethical responsibilities, and classic Western conceptions of democratic theory and practice place modes of communication at the very heart of the concept of civil society. Because this kind of continuing, mediated communication is so important in advancing national interests, democratic states have historically been keen to regulate broadcast media, to ensure that they don't fall into the hands of either vested private interests or state propagandists, and make certain they responsibly serve the high public purpose of informing, educating, and entertaining. In other words, governments, understanding the significance of the public space provided by media to the culture and practice of democracy, have sought to prevent damage to, or dilution of, that domain. It's what the Greeks called the agora; a place where citizens can gather to discuss issues of collective concern among themselves, away from the influence of state or commercial interests.

In this sense, all radio and television was at one time conceived of and organized as a public service enterprise rather than as just another industry. Most nations established state-financed broadcasters, which typically were given a monopoly. Where private broadcasting was allowed, it was invariably heavily regulated. Among the clearest statements of this ethic is contained in the landmark 1951 report of the Royal Commission on National Development in the Arts, Letters, and Sciences, known as the Massey Commission. The Commission was highly critical of submissions made on behalf of private broadcasters, who complained that the CBC, as a public broadcaster, represented unfair competition for their businesses. It noted that the private broadcasters appearing before it had failed to recognize that they had "any public responsibility beyond the provision of acceptable entertainment and community services" in their insistence that that they "must be left free to pursue their business enterprise subject only to limitations imposed by decency and good taste." The Commission report argued instead that:

> Broadcasting in Canada, in our view, is a public service directed and controlled in the public interest by a body responsible to Parliament. Private citizens are permitted to engage their capital and energies in this service, subject to the regulation of this body. That these citizens should enjoy adequate security or compensation for the actual monetary investment they are permitted to make, is apparent. But

> that they enjoy any vested right to engage in broadcasting as an industry, or that they have any status except as part of the national broadcasting system, is inadmissible.... they have no civil right to broadcast or any property rights in broadcasting. They have been granted in the national interest a privilege over their fellow-citizens. (238)

This attitude changed with the Reagan-Thatcher era of neo-liberal deregulation, which began in the 1980s and has continued. The assumption behind this new laissez-faire approach to media was that the freed-up market would better serve the public interest through the ineffable magic of Adam Smith's "invisible hand." The Americans were especially enthusiastic. The Reagan administration's Chairman of the Federal Communication Commission, Mark Fowler, and his aide Daniel Brenner in 1982 provided an explicit manifesto for neo-liberal ideology as it should be applied to public policy on broadcasting:

> Communications policy should be directed toward maximizing the services the public desires. Instead of defining public demand and specifying categories of programming to serve this demand, the [FCC] should rely on the broadcasters' ability to determine the wants of their audience through the normal mechanisms of the marketplace. The public's interest, then, defines the public interest. (207)

But it didn't turn out that way, and the so-called culture wars of recent decades has been one result of the massive free-market failures exposed by the withdrawal of statutory public service requirements in radio and television. The American media commons has been increasingly taken over by ideological bullies and shills for various commercial interests. On cable news and talk radio, adversarial debate has degenerated into hatemongering, which has been found to be profitable. And the content of mass media everywhere is increasingly designed to be a product for passive consumption, rather than a vehicle for communication in which audiences interact with and interpret what they listen to and watch.

For genuine communication to occur, the kind that causes change and adds to knowledge, people have to participate in culture in ways other than as consumers. And given the chance, they do, with enthusiasm. This was a bitter lesson learned by the big cable and telephone companies at the birth of the World Wide Web, which swept away their elaborate and hugely expensive preparations for the new, "interactive" world of digital media. Interactivity meant, for them, movies and other potted, proprietary content provided on demand. It was strictly a top-down conception, a straightforward extension of the existing one-to-many media structure which had been so profitable for so many years and which had conferred such enormous power on a handful of giant corporations. But the upstart World Wide Web gave users much more than that; it gave them the ability to create and distribute their own content, their own versions of common sense. They took to it with an alacrity that over a brief few months in 1994-95 wiped out the many corporate initiatives that were being pitched under the marketing phrase "the information superhighway."

"Cultural citizenship," a concept popular among media scholars, is the making of culture and identity by organic, bottom-up means. New digital media, and in particular blogs and social media like Facebook, Twitter, and a host of other personal communication applications, have given us all new means of playing a role in the creation of identity and meaning, and in shaping the conversation surrounding important social and political issues. But the very fact that these vehicles are disaggregated, fragmented, and highly personal, allows established for-profit mass media enterprises —

Big Media — to retain by far the loudest and most persistent voice in society. Think of advertising as the discourse of corporate interests, and then think of the ubiquity of advertising in our lives (on television alone, the average Canadian sees 25,000 commercials each year), and you will have some sense of corporate power and influence over culture and what we take to be common sense.[3]

THE PUBLIC BROADCASTER'S ROLE

Under siege, but still surviving as islands of social commitment in the sea of commercial, for-profit broadcasting are the original public broadcasters in most industrial nations, including our own. The Canadian Broadcasting Corporation (CBC) was created in 1936 on the model of the British Broadcasting Corporation (BBC), then, as now, the most respected broadcaster in the world. The purpose of the BBC, according to its founding director John Reith, was to "bring into the greatest possible number of homes... all that is best in every department of human knowledge, endeavour, and achievement." (221). Radio offered, for the first time, a way to address an entire nation simultaneously, and Reith and his generation were anxious to see it used for the highest purposes.

Contained in Reith's mission statement are the two most important defining characteristics of public service media: as near to *universal access* as is possible; and the *highest attainable quality* of content, in whatever genre, but especially in news and information.

Universal access is taken to mean that where you live, or how well educated you are, or how much you earn, should not lead to disenfranchisement. The logic of universal accessibility is not that of maximizing audience, but of equal rights. The issue of quality is more complex, and I will deal with in detail later on. However for the moment it's important to note that quality speaks to universality as well. For the programming output of a public broadcaster to be considered of high quality, it must be more than simply well made; it must appeal to a broad range of tastes and interests.

The British media scholar Michael Tracey has put it this way:

> Public broadcasting does not expect that it can please all of the people all of the time — indeed it sees in that approach precisely the kind of populism which nurtures cultural mediocrity, as quality is sacrificed on the altar of maximizing audience size. Public broadcasting does, however, believe that well-produced programs can please a lot of the people a lot of the time, and everybody some of the time. Public broadcasting is thus driven by the desire to make good programs popular and popular programs good: it understands that serving national diversity is not the same as "giving people what they want." (25)

Serving diversity means catering to tastes from popular to élite. One of public broadcasting's values is that, in doing so, it can hope to stir a latent interest or revive a dormant passion in some segment of the audience. It extends cultural horizons by offering audiences not only what they want, but what they may not yet know they want.

Public broadcasting is also uniquely in a position to provide programming for ethnic, cultural, gender, and minority interests, as well as for children. It is especially useful to the socially and physically disadvantaged. These are groups which almost by definition are neglected by commercial media, where audience maximization in affluent demographics is a crucial business imperative. Public broadcasting can both help them to communicate with one another and bring their interests to the attention of a wider public.

Public broadcasting has traditionally, as well, affirmed a commitment to education, in both formal and informal aspects. The CBC has for the past several decades neglected the formal approach, acknowledging that education is a provincial responsibility under the Canadian constitution. Education has been downloaded to provincial public service broadcasters, where they exist, and more recently the CBC has been expanding educational resources on its web portal, CBC.ca.[4] There is no counterpart on the CBC, for example, to the BBC-supported Open University, which offers about 600 courses leading to 250 formal academic qualifications, including post-graduate university degrees.

Finally, it is an axiom of public broadcasting that it should be free from the influence of vested interests, be they state or commercial. Programs paid for by advertisers inevitably have their content shaped by the business imperative of maximizing the audience and by the values inherent in consumer culture. On the other hand, programs directly financed by the state, without the intervening buffers afforded by institutional structures such as the autonomous Crown Corporation, will inevitably reflect political interests. He who pays the piper calls the tune.

But the independence required of true public broadcasting is more than simple insulation from political and commercial influences. It must also be morally independent. It must be free to explore, reflect, and criticize the values of both governments and commercial interests, and of communities to which it broadcasts. In this respect, the public broadcaster must be afforded the same kind of intellectual independence as a university.[5]

Canada's public broadcaster was given additional, special responsibilities by its founders that reflected the nation's unique geographic and political circumstances. It was conceived as a bulwark against the onslaught of American radio programming washing across the 49th parallel. And it was later to be given the additional responsibility of promoting "national unity," as in reducing the perennial tensions between anglophone and francophone Canada. The Broadcasting Act of 1932 provided for a single trans-national broadcaster that was to be fully bilingual right across the network.[6] Objections from English-speaking audiences, primarily in Ontario and the West, made this unworkable, and since 1941, there have effectively been two public broadcasters in the country, the English-language CBC and the francophone Société Radio-Canada or SRC. There has been surprisingly little cultural exchange between the two networks over the years, but what cross-fertilization there has been is virtually all there is in broadcasting in a country still remarkable for its two linguistic solitudes: private broadcasters have no interest in such worthy but unprofitable challenges.

Suspicions arose periodically that the French-language service was deliberately undermining national unity by promoting Quebec independence, and following the separatist Parti Québécois electoral victory in the province in 1976, Prime Minister Pierre Trudeau ordered the broadcast regulator, the Canadian Radio-Television, Telecommunications Commission (CRTC), to investigate. Another, similar, investigation was ordered by Prime Minister Jean Chrétien in 1995 following the whisker-thin victory for "No" forces in the Quebec referendum on sovereignty. In neither case was any evidence found that SRC had been engaged in propagandizing, or that its news coverage had been biased. What was noted was that SRC programming seldom reflected English-Canadian preoccupations, and the coverage of Quebec on the English network, though more complete, was less than adequate.[7]

In 1991 the Minister of Communications removed the CBC's statutory obligation to promote national unity, which was seen as a constraint on freedom of expression, and hopelessly incompatible with its responsibility to cover the news fairly and without political bias. The revised mandate required only that the CBC, in both languages, foster "a shared national consciousness and identity." This has been interpreted to mean the promotion of positive relations not just between French and English, but among all regions of the country.

In its original configuration, the CBC network offered, in addition to its own programming, an array of the most popular American radio programs. This had the dual purposes of weaning listeners away from American stations, while at the same time harvesting the substantial advertising revenue these programs supplied. (Most were sponsored by a single company or product; spot advertising was a rarity in early radio.) NBC and CBS hits like *Lux Radio Theater, Edgar Bergen and Charlie McCarthy, Amos 'n' Andy, Fibber McGee and Molly, Our Miss Brooks, Ma Perkins, The Guiding Light* and many more were already familiar to Canadian audiences who had been listening to them each night on powerful American border stations. The strategy initially increased the presence of American programming on Canada's airwaves, but it also helped to assemble an increasingly loyal listenership for Canadian programs like *The Happy Gang, Brave Voyage, John and Judy*, and the perennial *Hockey Night in Canada*, plus the many performing arts programs broadcast each week on the CBC network. That in turn fostered the development of a growing stable of Canadian performers, writers, directors, composers, and musicians while American content on the network was steadily reduced throughout the middle years of the twentieth century.

2

Public broadcasting under siege

The CBC has throughout its history been subject to more or less continuous attack from private broadcast interests and their lobbyists in Ottawa, who objected to competition from a publicly-financed institution.[8] Adding their voices to the chorus more recently have been neo-liberal ideologues who see state-subsidized enterprises of any kind as market-distorting abominations. A succession of timorous federal governments have for eighty years and more made mollifying compromises that have left the corporation in an ambiguous, confused, and ultimately untenable position.

The most obvious aspect of the CBC's crippling schizophrenia is its relationship to commercial sponsorship of programming. CBC radio was set free from commercial sponsors in 1975, but since its inception, CBC television (both English and French) has been financed in part by advertising, a source of revenue that becomes more important with each successive reduction in the federal government subsidy. Advertising currently provides roughly a third of the network's income, and is therefore critical to its survival.

With one foot in the altruistic ethic of public interest and the other in market values, CBC television has had to cope with conflicting mandates arising out of what media scholar Graham Murdock identifies as "a sharp distinction between those programs that people in their role as consumers of television most enjoy and those that, as citizens, they value for their contribution to the overall quality of public life" (192).

This distinction is reflected in two radically different notions of what constitutes the public interest. The market model thinks of serving the public interest as giving the public what it wants. A large audience is therefore a sign that the public interest is being served — otherwise those people would be doing something else. The theory further suggests that in a democratic society the state or its representatives have no right to make choices for citizens as to what they can watch or listen to, any more than what they read or write. Therefore, any definition of the public interest other than its own is anti-democratic.

The civic model, on the other hand, tends to put some distance between the audience and the programmers. It concerns itself with what the audience *ought* to be interested in, and not just what it has already demonstrated an interest in. It leaves space, in other words, for the audience to be surprised, and for taste and interests to be developed. Its focus is on the kind of programming that is intended to entertain and edify, but also to promote the public good by illuminating issues and promoting engaged and active cultural and political citizenship. The idea is that the state, in one way or another, should ensure that there is a depth, breadth, quality, and independence of media content that market forces alone will not supply. Of course there is nothing to say that programs cannot serve both market and civic interests simultaneously, as they sometimes do on CBC television, and frequently do on European public broadcasters.

There is something about the word "ought," as italicized in the paragraph above, which raises peoples' hackles when it's applied to the media. "Who the hell are you to tell me what I ought to be watching!?" Public broadcasting is often accused of being élitist because it dares to make such essentially moral judgments. But this is a misrepresentation, or at least an unfairly narrow interpretation, of what public broadcasting aims to do. It does not aim to force anything down peoples' throats; it strives instead to extend people's range of choices, and engage them willingly with content of a kind they are unlikely to find on any commercially-sponsored outlet.

And as to élitism, it is worth remembering that all radio and television programming, whether commercial or public-interest, is provided by one élite or another. Somebody else decides what you watch or listen to, and you have very little say in the matter. You can, in the jargon, vote with the remote, but your freedom to choose is confined to the menu presented to you in the TV and radio listings. And while the menu may be vast, multiplicity is no guarantee of diversity.

Who draws up the menu? Commercial media are programmed by an élite group of producers, mainly in Hollywood, who are highly skilled in deciding what will interest audiences and serve the needs of advertisers, while at the same time earning a substantial profit for the producers and distributors. They work in collaboration with advertisers and their representatives to ensure that commercials will integrate well with the overall package, and the right audience demographics will be addressed. Thus the choices offered by commercial media fall — *must* fall — within narrowly prescribed boundaries of what is compatible with sponsors' messages. That means they must not seriously conflict with the general ideological themes of consumerism and liberal market capitalism. And they must not be challenging or disturbing enough to make the advertisements that surround them seem crass or in poor taste. The commercial media élite, to the extent that they are successful, are highly skilled in navigating this landscape.

Public service media are programmed by a different kind of élite, by groups of men and women who are expected to devote themselves to serving the public interest. Like their colleagues employed by commercial media interests, they form an élite in the sense that they constitute a small, highly-specialized and skilled centre of power and influence. But they are, in principle, free to develop programming that stretches the imagination and challenges the intellect while entertaining and informing the audience. They can take creative risks no rational commercial broadcaster could countenance.

So the question becomes: which élite do you prefer to provide your media content? It needn't be one or the other; most of us like some of each. But few thoughtful people would choose the commercial élite exclusively.

THE BBC MODEL

The notion of public service broadcasting as it has been understood at the BBC, begins with John Reith's famous injunction, worth repeating here, that public radio should "bring into the greatest possible number of homes... all that is best in every department of human knowledge, endeavour, and achievement." The intent, a noble one, was Victorian in its paternalism. And when Reith spoke of a Gresham's Law operating in cultural affairs, in which the bad tends to drive out the good, he ascribed the phenomenon to the impoverished tastes and intellectual stolidity of the masses, who needed to be led to higher ground.[9] To educate and elevate tastes was the purpose of the BBC, and the reason why it had to be a monopoly: competition, and in particular commercial competition, would eventually result in broadcasters giving their audiences (who were in no position to know what was good for them) merely "what they want." Broadcasting, Reithian-style, was a benevolent project in top-down social engineering. His loyal successor as Director-General, William Haley, described the mission in a talk on the BBC Home Service in 1949 as "the disinterested search for the Truth," adding that "it should be frankly stated that to raise standards is one of the purposes for which the BBC counts." It must do this, however — and this was an important nuance — "within the broad contract that the listener must be entertained...[so that] while giving him the best of what he wants, it tries to lead him to want something better." (Tracey, 67)

And in a prescient comment that eerily foreshadows the current debate over our present state of media saturation, Haley insisted that broadcasting should be seen as an intermediary between the individual and the world of everyday experience, so that properly done, it would lead people to real-life experiences such as the theatre, or concert hall, or town hall meeting. The aim of broadcasting "must be to make people active, not passive, both in the fields of recreation and public affairs.... The wireless set or the TV receiver are only signposts on the way to a full life. That must finally lie in a sense of beauty and joy in all things, and in the experience of participating in life as a whole." (Tracey 68)

The challenge that preoccupied these early leaders of public broadcasting was expressed neatly by Harman Grisewood, a famous BBC actor, staff announcer, and eventually director of the high-brow Third Programme on the BBC: "How are we to ensure the continuity of our culture in an age of mass participation?" The culture he had in mind was one that had been described by Haley in 1946 as founded on "the things that matter...the ancient moral values" that "derive from Greece, Rome and the Holy Land...[forming] the basis of our civilization." (Tracey, 67)

This outlook strikes the contemporary Canadian's ear as quaint or even outrageous in its parochialism, as does Reith's paternalistic authoritarianism. And as British society went through the wrenching post-war transformations brought about by massive immigration, the rise of consumerism, the market's unrelenting preaching of egoistic individualism, and the profound skepticism of newly fashionable continental philosophy, it challenged the old model. The public broadcaster's function might still be to serve the public good, but it was no longer clear what that good looked like. For better or worse, clarity about the very concept of good was one of the casualties of post-modernism.

Throughout the 1950s and 1960s, the BBC was faced with new social realities and with a rising tide of commercial, for-profit competition first from Europe and then at home. It struggled to find the sweet spot between élite and populist tastes. A 1957 internal inquiry into how the corporation ought to respond acknowledged that, "the loss of its monopoly in broadcasting has very much reduced the BBC's power to manipulate its program policy in the interest of social and cultural aims," adding: "It is instead engaged in a battle for its position as the nation's home entertainer, a position it must retain if it is to continue as the mirror of the nation's great events and a cultural and educational influence of social importance." Populism had effectively gained the upper hand. The BBC would henceforth be obliged to service "its many audiences more as it finds them than as it would wish them to be" (Tracey, 82). The question now became whether or not that need necessarily led to a lowering of standards or a betrayal of the original public purpose and high ideals.

The internal logic of the continuing renewal process at the BBC would eventually produce one of the corporation's most successful Director-Generals in Hugh Greene. His concept for the BBC rested on the need to continue its traditional mode of financing through a licence fee on radio and television sets, in order to maintain both its arms-length relationship to government and its commercial-free programming schedule. Even minimal commercial financing would, he said, be a slippery slope: "If we were successful in the commercial field there would be inevitable political pressure to deprive us of our licence revenue, gradually but in the end totally." (Tracey, 85) This has certainly been the experience with the CBC, where the success of television advertising has provided a rationale for freezing the annual subsidy, and has led to repeated calls for the re-introduction of advertising on radio.

Commercial broadcasting, Greene said, was characterized by salesmanship rather than service, "the mere desire to attract attention" by any means. "I believe the profit-servicing system is defective. Its interests can never be towards providing the best in every category — its concern is fundamentally with sales and its categories of broadcasting are categories of salesmanship." Not only were commercial broadcasters obliged to tailor programs to suit advertisers, their popularity as measured by ratings, "is bound to be a deceptive and specious one." Why? Because the requirements of public service, he said,

> ...are as manifold and diverse as the individuals who compose the public. A library considered as a public service could hardly be correctly evaluated merely in terms of the number of books borrowed. If this were the standard of measurement, libraries would have a very easy road, and we could easily guess at what the contents of this library would be. But a public service of broadcasting, like the library, must provide so far as possible for every taste and for every sort of entertainment, for information upon every worthwhile topic, and for education wherever it is needed. (Tracey, 88)

Furthermore, whatever the public broadcaster did, Greene insisted, must be informed by certain values. If that was interpreted as paternalism, then so be it. He said in a speech to the National Association of Adult Education in 1961: "I hope I shall not sound 'undemocratic' if I say that by and large it is fairly well agreed in our society that knowledge is better than ignorance, tolerance than intolerance, and active concern for the arts or public affairs better than indifference, and that wide interests are better than narrow."

THE NEW CBC

Greene's attitude to commercialization and to addressing audience tastes provides a stark contrast to the approach to advertising and programming taken by the CBC during the transformative, six-year reign of Richard Stursberg as head of English-language programming, from 2004 to 2010. In 2012 Stursberg published a remarkable tell-all memoir of this turbulent vice-presidency, one that many students of public service media and the CBC have found stunningly perverse in its conception of, and attitude toward, public service broadcasting. It was wittily, if derisively, entitled *Tower of Babble*.

Stursberg was, if nothing else, a champion of catering to populist tastes. During his tenure, talk of public service mandates was literally banned. "We did not want to produce university lectures, books or performing arts.... We wanted to work within the television conventions that English Canadians preferred. We would jettison 'edgy,' auteur-driven projects for season-long series working within understood narrative traditions. We would make police procedurals, situation comedies, reality eliminations, lifestyle shows, and quiz shows. We wanted to make TV for the largest possible audience." (Stursberg, 80)

One of his first actions in developing a new programming schedule was to cancel the multi-award-winning *Opening Night*, two hours of prime time, advertising-free performance spectacles each Thursday evening. "There were plays, ballet, one-person shows — beautiful performances. The shows were made to a high standard. The audiences were dismal...the CBC could not afford to give up the most important [ratings] evening of the week to a show that generated no revenue and rarely made 200,000 viewers. Whether we liked it or not, and despite how beautiful the shows were, Canadians were saying no." (Stursberg, 81)

Contrast that with comments made in *The Times* by an embattled BBC Director-General Alasdair Milne in 1981, as Thatcherite Conservatives clamoured for the dismantling of the public broadcaster:

> The basic premise of public service broadcasting, as I understand it, is this: if you address yourself to the nation as a whole, you must appeal to the nation as a whole — in

> all its diversity. All of us have amazing varieties of tastes, interests and curiosities. Each one of us belongs, at one and the same time, to majorities and minorities. What public service broadcasting must constantly seek to do is to provide enough satisfaction in the belief that allegiance to taste and interest is never certain, is constantly changing, and that therefore you must offer the widest variety of programming. (Tracey, 111)

The difference between the two points of view here — and "difference" seems far too mild a term — is that while Greene and Milne wanted the BBC, to the extent feasible, to serve all tastes and interests, Stursberg was content to cater to only the most populist. He was openly disdainful of anything else, believing that television failed whenever it strayed from the conventions of American network programming. "Television is fundamentally about entertainment. It is the medium par excellence that people consume to be told stories, to be made to laugh, to be thrilled, frightened, moved, charmed or excited." Therefore, to "suggest that the CBC should make shows that were alien to what Canadians knew and liked seemed almost perverse." (Stursberg,18)

What he apparently did not understand was that those conventions, so familiar to Canadians and developed over decades of experimentation, had nothing to do with the nature of television as a medium and everything to do with the needs of advertisers to keep audiences in an appropriate frame of mind to tolerate and absorb commercials. He was simply wrong in his conviction that television is fundamentally incapable of taking audiences beyond Hollywood orthodoxy. How he sustained this belief in the face of spectacular contradictory evidence produced by the world's public service broadcasters, much of it available to Canadian audiences on outlets like PBS and BBC-Canada, is a mystery. [10]

Nor are audiences naturally or innately predisposed to love narrative, Hollywood-style. It is their preference because it is what they have become accustomed to. The television audience in North America has for decades been purposefully groomed to accept the programs they watch in the absence of fare that might be more challenging, or disturbing, or thoughtful, or inspiring, or locally relevant. Programming that falls into any of these categories is unlikely to find corporate sponsors, because it provides a poor environment for their commercial messages. And serving advertisers is the raison d'être of commercial broadcasters. In the stacks of any university library are shelves of volumes on precisely the problem of Hollywood standards overwhelming, and ultimately changing, local cultural preferences by sheer weight of exposure. It is a theme central to the study of globalization.

Unlike the BBC's Greene (and his predecessors and successors without exception), Stursberg apparently saw nothing wrong with commercial sponsorship on public television. He never questions its appropriateness in his memoir. In fact, he prides himself on having encouraged the CBC's sales department to promote the public broadcaster as the place to go, not just for conventional spot advertising, but for the more devious rewards of product placement — advertising by stealth. In one of the more astonishing passages of his book he writes:

> In thinking about revenues we knew we had certain advantages over our competitors at CTV and Global. Whereas they bought most of their shows ready-made in the United States, we commissioned or produced most of ours from scratch, which meant that we could incorporate advertisers and sponsors directly into the shows as they were being produced. We could do product placements, website extensions or games and contests that were integral to the

> shows themselves and helped to sell the sponsors' products. Advertisers were happy to pay a premium for that, and our competitors could not match it. (247)

Some would see this as a stunning betrayal of the public trust. In what conceivable interpretation of the public interest is it acceptable for a public broadcaster to offer up its programs to advertisers for purposes of selling their products subliminally, by stealth?

But Stursberg is oblivious: "One of the best examples of how this could be done was *Kraft Hockeyville*," he goes on to boast. *Hockeyville* was a heavily-promoted reality show concocted by Stursberg's programmers in which small town Canada competed for corporate-donated prizes by affirming their devotion to hockey — a sport which, not incidentally, fills about 350 very lucrative hours on CBC television each year. Stursberg continues: "In the show and the online voting, Kraft was woven into everything. *Hockeyville* appeared on Kraft products in supermarkets, and everybody was pleased. To extend the idea further, we commissioned an episode of *Little Mosque on the Prairie* for Kraft.... In this particular episode we had Mercy [the fictional Saskatchewan town in which *Little Mosque* is set] apply to be named *Hockeyville*." He concludes his story: "In 2008 and 2009 *Marketing* magazine named CBC "Media Player of the Year" for its innovations. That had never happened before." (247-48) *Little Mosque* featured product placement in a deal with TD Bank in 2009 (a character visits a branch to see if he has enough money to throw a party), and in *Being Erica* a lead character was given the job of manager of a TD bank branch. The bank also appeared by stealth in *Heartland*. Worse, in 2010 CBC teamed up with Kellogg to promote Eggo Waffles in a segment on healthy breakfasts with the CBC puppet character Mamma Yamma.[11] Since the departure of Stursberg the corporation has stopped announcing its product placement deals in press releases.

(An aside, for context: paid product placement was permitted for the first time on British commercial television by the industry regulator, Ofcom, only in 2011. At the BBC, the ban continues. Under the new regulation, commercial stations must display the letter "P" at the top right corner of the screen for three seconds at the beginning and end of a program that contains product placement. As well, Ofcom ruled, the placement must not interfere with the editorial content of the program, nor be too prominent. And a total ban remains in place for children's programs, news and current affairs programs, and alcohol, tobacco, medicines, escort services and products that are high in sugar, salt and fat!)

And then there is this. In response to the financial crunch that faced the CBC in 2009, brought on by the economic slump and resultant drop-off in advertising revenue, Stursberg proposed that, "we should jettison the block of pre-school shows for children in the mornings." Never mind that the CBC, like all other public broadcasters, considered children's programming to be absolutely central to the fulfillment of its public service mandate. Never mind that the quality of CBC's children's programming had earned a world-wide reputation, and programs like *The Friendly Giant* had imprinted themselves on the Canadian psyche. Stursberg had noticed that they "carried no advertising and had very small audiences" (262). At the same time he asked the CBC Board of Directors to reverse the corporation's long-standing policies banning infomercials and political advertising outside election periods. In both cases he was turned down. The board felt that infomercials would erode the network's credibility and gravitas, and that political ads would undermine the CBC's perceived independence (251-52).

It would be difficult to imagine better illustrations of why a reliance on revenue from commercial sponsorship, and the consequent focus on audience numbers, are a danger to the public service responsibilities of public broadcasters.

The argument is sometimes made that, for a public broadcaster, a little advertising is acceptable, so long as it doesn't become a dominant source of revenue and thereby begin to shape programming through the drive to achieve higher ratings and higher income. The examples of Britain's ITV and, more recently, its spinoff Channel 4 are often proffered. While it is true that the UK's first commercial channels produced and commissioned some great programming (*Inspector Morse, Prime Suspect, Jewel in the Crown* and many others) it is also true that they have operated since their inception under a regulatory framework that insists they pay essentially as much attention to quality and public service values as the BBC.[12] In Canada, no such regulation exists for commercial broadcasters, nor is it even conceivable that it could be introduced at this late date.

So the dilemma remains. As long as CBC television is forced to rely on commercial sponsorship for a substantial portion of its operating income, there will be powerful temptation for it to produce programming designed primarily to grab ratings. And, of course, while the BBC's Greene was aware that the library that judged its success solely by the number of books borrowed would be a poor library indeed, Stursberg, and by all appearances his successors, energetically embrace the equivalency of numerical popularity and success. "If not ratings, then what?" Stursberg wants to know. It's a question I'll try to answer in the following chapters.

3

"Ask the horse!" The trouble with ratings

Richard Stursberg, who engineered a cultural coup within CBC television during his tumultuous six-year tenure as Executive Vice-President, English services, was dismissed by the corporation's President Hubert Lacroix in 2010.[13] His legacy lives on, however, and since that time, he has continued to promote his vision of a CBC that leads the ratings in populist programming at the expense of every other genre, claiming that, thanks to his visionary strategy, "it is possible to say with certainty that the CBC has never been stronger." Writing in *The Globe and Mail* in September 2012, Stursberg claimed that, thanks to his leadership, "For the first time in history, the CBC has proven that Canadians can make entertainment shows that can compete with the programs made in the United States... If there ever was a Golden Age for the CBC, it is now."[14]

On the other hand there are many Canadians, including legions of loyal CBC radio addicts, who see in the television service what *Globe and Mail* television critic John Doyle sees: "a blinding sheen of lightweight nonsense." A schedule, punctuated by ratings-grabbing gimmicks like *Battle of the Blades*, in which "there isn't a single serious-minded cable-quality drama...a single searing comedy...nothing to compel anyone to note that no other broadcaster would air such a program." (Sept. 10, 2012)

The strongest journalistic resources in the country are housed in the CBC, shared among radio, television and the cable news network. But they are often poorly displayed in the grossly overproduced *The National*, which Doyle harpoons in the same column as "sometimes a disgrace, a meandering journey through the mind of a flibbertigibbet who spent the day garnering news bits from a hodgepodge of online sources." What can possibly account for such widely divergent views of CBC television? The answer lies, first of all, in two starkly different definitions of success. Doyle argues that, "The CBC is mandated to be more than a broadcaster. It is mandated to be a cultural institution, an incubator of artistic talent, employer of talent from many genres and provider of unique programming that other broadcasters fail to deliver." For Doyle, the CBC is mandated to broaden public taste, rather than pander to it — to provide a venue for excellence. Success is, or ought to be, measured in that context.

For Stursberg, success for the CBC has a different meaning. As he makes clear in his memoir, television is not a medium for high artistic expression, but rather for mass entertainment. He subscribes to the attitude, popular in the commercial media industry, that the success of a television offering ought to be measured instrumentally, in the utilitarian arithmetic of ratings, and that a discussion of quality that does not take place in the context of popularity is simply specious. Viewers choose to watch elimination reality TV shows in large numbers because it is quality television: how do we know it is quality television? Because viewers watch in large numbers.

There is nothing subtle about Stursberg's view on this. In describing his strategy for renovating CBC television in *Tower of Babble* he says there would be only one measure for success: audiences. "It would be a brutal standard.... It would no longer allow [the corporation] to fudge the meaning of success by talking vaguely about 'mandates,' and 'quality.' It would be a standard by which shows, producers, stars and executives would be judged." (23)

At first blush, his approach to measuring success for CBC television seems eminently sensible. As he himself asks, "If not audiences, then what?" Hence his obsessive focus on ratings.

And indeed, if it were truly audiences that were being measured in the television ratings services in which he places his faith and to which he offloads responsibility for quality evaluation, then it would be harder to argue with him. But here's the thing — ratings do *not* really measure audiences in any reasonable or reliable fashion. They provide little or no information as to whether viewers were bored or entranced, stimulated or maddened, inspired or nauseated, made joyful or angry, educated or misled. In short, they provide no clue as to whether or not the public interest has been served. They simply provide an indication of how many warm bodies were located in front of television sets tuned to a particular item on the menu of programs placed on offer by the TV industry.

Audiences are composed of real people. Ratings, on the other hand, are a completely artificial statistical construct designed to do the impossible — to make a homogeneous, packageable commodity out of millions of individual consumers whose tastes, needs, and responses are unique, and who are as diverse as their numbers. Ratings are to audiences as hamburger is to an Angus steer.

Ratings, in short, say nothing coherent about whether the audience is being well served. Then why do they exist? Because markets, which are the means through which commodities are distributed in an economy, run on numbers. In the peculiar market that is commercial television, the commodity being sold by broadcasters is *viewers for commercials*. In order to turn those individual viewers into a packaged, uniform commodity that can be served up to potential purchasers — i.e., advertisers — broadcasters have turned to the ratings services, which attempt to measure total numbers of viewers for any program according to various criteria, including age and income bracket. Ratings provide the "empirical evidence" that broadcasters need if they are to demonstrate to advertisers that their money is well spent. In my own experience in the industry, it has always been understood that ratings provide a largely spurious metric, and the main concern among industry managers is that the ratings services treat all competing broadcasters equally within this fictitious construct. They insist on a level playing field.

Whether the ratings numbers are accurate in terms of volume (and this is often disputed) is of less importance to the industry than whether they are collected fairly and impartially. This is because ratings are a kind of currency, like paper money, and currencies only work if everyone agrees on their underlying value. The currency of ratings allows a broadcaster to say to an advertiser, "I will give you an audience of 450,000 at 6 p.m. each weekday in return for our rate card fee of so much per thousand viewers." Another broadcaster might then undercut that offer by serving up higher audience numbers, or a lower fee, or both. This is competition at work, doing its job, with ratings providing the necessary numerical data.

Implicit in the sales pitch for audiences is the promise that they are composed of happy campers; no sponsor wants to purchase a disgruntled audience. This guarantee is provided by the commercial television industry's oddly illogical, but universally accepted formula for audience satisfaction. It works this way: "Why do audiences watch our programs? Because they like them. How do we know they like them? Because they watch them." On this tissue of sophistry rests the prime axiom of commercial broadcasting, which is that, "We give the audience what it wants." Even more perniciously, this axiom is extended into the realm of quality: "Why do audiences watch our programs? Because they're good. How do we know they're good? Because audiences watch them. Ratings tell us so." [15]

In the real world of network television, on the list of variables that actually determine a program's audience ratings, "quality" often ranks relatively low. It may be less significant than such key determinants as the program's place in the broadcast schedule, the popularity of its lead-in and lead-out programs, what's on competing channels, marketing expenditure, and so on. It is part of the logic of commercial broadcasting that this is as it should be, and that any attempt to make judgments about program quality in broadcasting is bound to be authoritarian and oppressive, or at least ill-advised and counter-productive — an unwarranted interference with the transcendent moral authority of the free market. It is part of the logic of commercial broadcasting that audiences should, and do, make judgments of quality for themselves, through their choices in the free, self-disciplining market. It is an iron-clad principle celebrated as "consumer sovereignty."

This hands-off, relativistic, attitude to quality ("the market will handle that") finds favour among industry executives because it lets them avoid difficult moral and aesthetic issues that cluster around the idea of quality. And that's a relief for them, because engaging with quality might well result in program expenditures not directly related to audience size and advertising income. In other words, wasted money and lower profits.

It is a fundamental tenet of capitalism that competitive markets contrive to produce products of the highest attainable quality at the lowest feasible price. The basic or idealized explanation for this is that an entrepreneur who offers a product or service that proves to be inferior to, or more expensive than, one offered by a competitor down the street will lose business to that competitor. Consumer sovereignty rules, and consumers (who are assumed to be rational) will choose the better value. This in turn is a powerful incentive for the uncompetitive business to improve its product or lower its price, or both, if it wants to survive. In this way, overall product quality ratchets up, as price trends downward. We see this at work every day in markets like consumer electronics, where a tangible product is being manufactured and sold. In competitive markets, all things being equal, *the good drives out the bad.*

But when this logic is applied to the commercial media market, e.g., the market that produces television programs, it breaks down. A paradox arises. In the highly competitive television market, even though consumer sovereignty is well-served, an entrepreneur's success tends to be an indicator of *lower*, and not higher, quality in programming. Product quality will indeed rise over time thanks to market dynamics — but the product in question is "audiences." What is a "high quality" audience in the eyes of the advertisers, who are its potential purchasers and consumers? The quality audience has three characteristics: (1) the correct demographics; (2) the largest numbers; (3) the lowest price. Dramas, sitcoms, newscasts, and reality shows are what broadcasters offer up to pull together the audiences desired by advertisers; the less money they spend in that process — in program production — the better, for both broadcaster and advertiser. And so, in a competitive market, the investment made by broadcasters to accumulate a given audience will be forced inexorably downward. And because there is an obvious, direct relationship between the cost of programs and their quality (again, all other things being equal), program quality will necessarily follow the same downward trend.

This happens not because of some flaw in human nature in general, or in the character of media managers in particular (though these may exist). The question, "Left to its own designs does popular culture gravitate towards the laudable or the dire?" asked by many a media scholar, is the wrong question when it comes to commercially-sponsored media. What happens to quality in commercial media is precisely that popular culture is *not* left to its own designs. What happens is the result of the mechanical operations of the unregulated market, combined with the supremely rational behaviour of large corporations. Whatever they might once have been, by the late twentieth century these entities had come to see their sacred responsibility as maximizing the value of the assets under their control, on behalf of corporate shareholders. The quality of programming is, in principle, irrelevant to this goal, so long as it is within legal and regulatory bounds, and therefore cannot expose the corporation to legal penalties. As CBS programming executive Arnold Becker asserted, "I'm not interested in culture. I'm not interested in pro-social values. I have only one interest. That's whether people watch the program. That's my definition of good, that's my definition of bad." [16]

So economic success for commercial broadcasters operating in a competitive market implies, as a necessary corollary, mediocre programming. This is Gresham's Law of commercial media: where media are supported by advertising in a competitive market, *the bad drives out the good.*

The proof: critics reel back in astonishment on the rare occasions when an anomaly occurs in the form of great TV on a commercial network. These exceptions prove the rule, as do genuinely high-quality television programs routinely turned out by broadcasters and producers that do not rely on commercial sponsorship, such as subscriber-supported HBO. Subscriber-financed broadcasters and producers operate in a market in which programs (and not audiences) are the product, and to the extent that classic market dynamics affect them, they tend to force program quality upward, rather than down. Public broadcasters tend to operate entirely outside that market, and standards are imposed internally, by producers themselves.

When *The Guardian* newspaper asked its gaggle of arts and television reviewers to pick their all-time best TV dramas, the list, which was topped by *The Sopranos*, was overwhelmingly composed of public broadcasting productions (mainly BBC) and American subscription channel productions, mostly HBO.[17] While subscription channels do rely on audience popularity (as expressed in subscriber revenue) for their survival, they have much more freedom to experiment, and to serve niche audiences, than do commercial broadcasters. Their programs are designed to appeal to existing and potential subscribers — i.e. actual viewers — rather than to advertisers, and thus the worst of the dynamics of Gresham's Law are avoided. Program budgets are shaped by a desire to expand subscription numbers, rather than to supply low-cost solutions for advertisers. Expanding subscription numbers will frequently mean pushing boundaries and taking risks, exactly what risk-averse commercial broadcasters and their advertiser clients try to avoid.

Before moving on to the topic of quality itself, it needs to be asked whether advertising per se, removed from any context of competitive market dynamics, is a corrupting influence on programming. In other words, might it be possible for a public broadcasting system like the CBC to accept advertising and still maintain the highest standards of program quality, as defined within the context of a public service mandate? After all, it's the competitive pressure to produce audiences at lower and lower cost that drives quality down in the purely commercial market. If that dynamic is removed, would it not be possible to maintain program standards and still benefit from advertising revenue?A compelling account of the impact of advertising on CBC television has been provided by one of Canadian public service broadcasting's founders and one of its most reliable historians, W. Austin Weir. Writing in 1964, he said:

> There has undoubtedly been a major slide toward commercialism within the CBC.... [T]he urgent need for money to keep the ever-expanding machine going has brought a high degree of concentration on commercials, a condition extremely difficult to resist.... Today the CBC's sales organization...has adopted all the paraphernalia of big business and thrives on charts, curves, and sales targets, a system of sales incentives, and finally commissions to all salesmen exceeding set targets which vary for the several departments.... More and more the thinking has had to

> be what will please sponsors, what will get maximum audiences, what will sell.... *Commercial pressures are natural, persistent, inexorable, and those who have never been in the business have no idea how insidious and compelling they can be in the face of tightening budgets*. (313; emphasis added.)

While there is a theoretical argument to be made that public service could co-exist successfully with commercial sponsorship, on-the-ground realities militate powerfully against it. The logic of neo-liberalism shows no sign of losing its grip on politicians of all stripes, and within that logic, the public broadcaster that relies on advertising revenue finds itself on a slippery slope that will lead over time to lower public subsidies (always a desirable outcome) and therefore to yet greater need for commercial sponsorship (also desirable, because it reduces the need for public subsidy while serving commercial interests). After all, according to Grover Norquist, Washington lobbyist and author of the American "Taxpayer Protection Pledge," the ultimate goal of neo-liberalism is government "so small that it can be drowned in a bathtub."

Moreover, from the point of view of modern management culture, efficient administration of any hybrid public service/commercial broadcasting system such as the CBC demands that advertising revenues be maximized. Allowing potential revenue to go uncollected is the very definition of management irresponsibility. It's also an enormous political liability when it comes time to negotiate public subsidies. While the culture of public service broadcasting ought to be completely different from the culture of for-profit commercial broadcasting, as a practical matter the intrusion of the advertising economy makes that distinction difficult to achieve and maintain. [18] As does the neo-liberal assumption that the administrative methods of private industry are always superior to those of public service enterprises.

Not all arguments against advertising in broadcast media are political or economic. Advertising intrudes into content in ways both overt and subtle. In the first case, as I've already noted, advertisers typically do not want their products to be associated with controversial programming or programming of a serious nature in which commercial breaks are an obvious and annoying distraction from engaging content; thus, many a brilliant program idea is smothered before ever seeing the light of day. But the very presence of advertising breaks in a program schedule shapes the structure and content of the programs themselves. In contrasting the programming of Britain's hybrid commercial/public service ITV network with that of BBC television, media scholar and television producer Richard Rudin notes that the public broadcaster's advertising-free schedule was able to adapt a program's length to the requirements of its content (for example, talk shows could be extended by five or ten minutes if necessary to adequately explore an idea, or a drama might need an extra five minutes to complete a satisfying narrative arc). But commercial broadcasters,

> ...have to be virtually obsessive in keeping to the times scheduled and paid for by advertisers. These breaks are now computer-controlled and "live" output such as rolling news channels have to structure their output round the advertising breaks.... An ITV comedy, soap opera or drama not only had to fit a commercial half-hour (usually about 26 minutes of actual program time) but had to have a narrative "hook" to keep the audience through the mid-way commercial break. In the original UK legislation commercials were confined to "natural breaks" in programs, e.g. between scenes in the narrative, so as not to be too intrusive on the action. In an hour's drama these would generally be around the 20- and 40-minute mark. However the easing of restrictions on both commercial minutage and the timing of commercial breaks in the twenty-first century resulted in the introduction of an additional fourth break per hour. So, when dramas of an earlier period are repeated, there is often a jarring cut of scenes into and out of the breaks. (Rudin, 129)

In North America, of course, this has always been the reality, to the degree that it is rare for a program other than a sports broadcast to run beyond its scheduled length.

In US television programs, the convention has long been to have a break following an initial scene in a comedy or drama, before the opening titles, so the first scene had to build up suspense or comedic tension to lead the viewer through the commercials and be hooked into the program. Programs made in the UK, but with the US market in mind, adopted this technique, even though in Britain there would be no commercial break. Similarly, US dramas and comedies would have a break before the final scene and "end titles," so there had to be a "pay-off" scene after the main drama had been resolved to keep the viewer tuned through the commercials.

There is indeed a Gresham's Law operating in commercial television, but it is driven not by any implicit preference among the "masses" for junk. It is a feature of the dynamics of the commercial media market in which producers compete for advertisers by selling audiences. Richard Stursberg amusingly, and disastrously, misunderstands this when he writes in his memoir that under his administration and its new, populist programming strategy: "There would be only one measure of success: audiences. If Canadians did not watch, it meant they did not care.... In the words of the BBC, whose slogans are plastered all over the walls of White City, their great studio complex in west London, 'Audiences mean everything to us." (23)

Of course, what is meant by this slogan is not, "Ratings mean everything to us." It means, in fact, exactly the opposite: "Service to our audiences is what we are all about."

4

"If not ratings, then what?" Defining quality in public broadcasting

Okay, so ratings don't measure program quality. What does? If not ratings, then what? How is it that critics like the *Globe*'s Doyle and his colleagues can claim to know quality when they see it and criticize the CBC TV for not producing enough of it? What did Reith have in mind when he spoke of bringing only the "best" to the BBC's audiences? For that matter, what do commercial broadcasters mean when they make distinctions between good and bad programs? What does Stursberg mean? What are we talking about when we talk about quality in media production?

In his memoir, Stursberg explains that, before he instituted his ratings-grabbing populist program strategy at CBC television, "there were no objective performance standards" in place at the corporation. Success was measured not by ratings, but in terms of "vague notions" of public service and distinctiveness. Never one to let logic stand in the way of an argument, he interprets this as meaning that, prior to his arrival, people at the CBC presumed that if a program was popular, on their own network or a competitor's, "it must be vulgar and stupid." But, he insists, the "most popular American shows that Canadians watched in their millions were not poorly made rubbish. To the contrary, they were often beautifully realized, well written, well acted, well directed and well produced. The top shows — *Law & Order, CSI, Desperate Housewives* — were exceptionally good by any standards." (11)

By any standards. We can be charitable here and assume that he does not mean this to be taken literally. What he apparently means is that in terms of the quality of the craftsmanship involved in making these shows, they were exceptionally good. They are well written, acted, directed, and produced examples of the craft of television-making. His protestations to the contrary, he would have had to search long and hard at the CBC to find anyone who would disagree with this. What he might have found, however, were many committed public broadcasters who would deny that artifice automatically results in good quality. To use an obvious example, the film director Leni Riefenstahl's 1935 paean to Nazism, *Triumph of the Will*, has been widely recognized as a technical triumph — beautifully shot, produced, edited, and so on — but utterly meretricious. (Readers will have their own contemporary examples; mine would include *Pulp Fiction*, *Kill Bill* and other Quentin Tarantino films.)

In order for a program to be characterized as good in the everyday sense of that word, it clearly must have more going for it than technical excellence. This is particularly true in the context of public service broadcasting, which sets its sights higher than simply amusing or entertaining its audience. "Television is good," says media scholar and consultant Geoff Mulgan, "when it creates the conditions for people to participate actively in a community; when it provides them with the truest possible information; when it encourages membership and activity rather than passivity and alienation; and when it serves to act as an invigorator of the democratic process rather than as a medium for what Walter Lippmann described as the manufacture of consent" (23) "Exceptionally good" public service programming aims to be the best it can be technically. But it must also contribute to what might be called the moral economy of the nation. It should in some way, large or small, leave viewers better than they were. Better, more informed and involved citizens; better, more fulfilled human beings. I leave it to the reader to judge whether *Law & Order, CSI,* and *Desperate Housewives* are "exceptionally good" in this sense.

Some will object that TV is often used as a simple pacifier, a relaxation device and stress-reliever, an assist to "vegging out" after a hard day. That, too, is good TV because it serves a useful purpose, they will argue. And no doubt there is some truth to this. But people watch television with varying levels of attentiveness; they may even fall asleep! There is no reason I can think of why a good program in a popular genre, one that may be engaging to an alert viewer, should not also allow viewers to unwind.

In a desperate and, in retrospect, misguided attempt to achieve a ratings turnaround at CBC television, Stursberg and his programmers poured the network's dwindling resources into making television shows that drew audience ratings that were occasionally comparable to the mostly American content on the private networks. The programs, from police procedurals to reality shows (the CBC euphemistically calls the latter "factual entertainment") are virtually indistinguishable from their American phenotypes in every respect except setting. The question this strategy was supposed to answer was one asked by both the ill-informed and the willfully perverse among the public broadcaster's critics: " How can you justify spending all those millions of taxpayer dollars if your ratings are so dismal?"

In other words, in responding to the jibe about inferior ratings, the CBC accepted the logic it's based on. But the logic of the equation: quality = high ratings, is false, from a number of points of view. To reiterate:

- Audiences are able to select their viewing choices only from the menu that is placed before them by broadcasters in their broadcast schedules. What's offered in the schedule is determined, in the case of advertising-supported broadcasters, by advertisers, not audiences. To the extent, then, that audiences are able to "vote" for their preferences by watching, they are picking and choosing among an artificially (and arbitrarily) restricted selection. *What they would really prefer to be watching might well be "none of the above."*

- The level of consumption of a product says nothing about its quality, about whether it's good or bad. It speaks rather to such considerations as price, availability, the absence of options, marketing cleverness, and any number of cultural considerations such as fads and peer pressure. Just as quality is not always rewarded by popularity, popularity is not always a sign of quality.Even conceding, for the sake of argument, that high ratings can be a reliable indicator of genuine popular demand for a program or genre of programming, it still does not follow that "high ratings = high quality." It is possible, even routine, for people to sometimes desire that which is not desir*able* — not a worthy object of desire, or more concisely, not good. And this can mean "not good" for the individual in question, or not good socially, or not good in a more abstract, normative sense, or all three.

So we're back to the original question: what are we talking about when we talk about quality in television programming? If the goal of public broadcasting is to provide quality content, how do we measure success or failure? If not ratings, then what?

The puzzle has received some attention from communications scholars in recent years. The hope is that an approach can be developed that will allow a broad consensus to be formed around quality judgments. One way they have approached the problem is to break the concept into categories: sender-use quality; receiver-use quality; craft (or professional) quality; and descriptive (or truth) quality.

Sender-use quality is judged according to how well programming fulfils the needs and desires of the broadcaster. In the case of a commercial broadcaster, that will boil down to how much money the program makes. It's not quite so simple in the case of a public broadcaster like the CBC, where determining quality will involve such criteria as conformity to the broadcaster's official mandate and more generally to a responsibility to inform, educate, and entertain a national audience.

Receiver-use quality is quality in the eye of the beholder: as a subjective measurement it can be and often is completely idiosyncratic. One viewer's hockey bliss is another's outrage at violence on ice; even the most brilliant documentaries seldom have broad appeal, but are intensely enjoyed by self-selecting audiences. Moreover, audiences of different cultural backgrounds may respond differently to the same program, taking away different meanings.

Craft quality is what gets rewarded by prizes that are refereed by industry practitioners. It is the expression of the hard-won skills of the media professionals who make up the cast and crew of any production. It is the quality that, along with talent and inspiration, makes a screenplay or a musical performance great.

The fourth category, descriptive quality, is the classification most likely to cause serious debate. The question that has to be answered is: how close does this program come to portraying the world as it really is? Does the program, whatever its genre, make an honest, sincere, and responsible attempt to present the truth about the world?

Broken down in this way, the task of measuring quality in programming becomes less daunting, and entirely removed from the brutal, reductionist realm of ratings. It's possible to imagine a continuing conversation between programmers and the public that could arrive at a kind of evolving consensus. It becomes possible to answer the question: "If not ratings, then what?"

Another approach is to turn for answers to the people who produce television for a living, the artists and artisans who make a profession of television programming of all kinds. Research into professional attitudes to determining quality, gleaned through dozens of interviews and summarized by media scholar Irene Costera Meijer boils the issue down to the following checklist:

- Were the craft skills that went into the making of the program of a high standard?
- Was the program adequately resourced?
- Was it serious and truthful?
- Was it relevant to the concerns of the day?
- Did the storytelling touch the emotions?
- Did it appeal to curiosity/provoke thought?
- Did the program-maker have a clear objective? And push to achieve it?
- Did the program–maker have a passion/commitment that gave energy to the program?
- Was the program innovative, original, or adventurous?
- How did the audience react to it — in appreciation as well as numbers? (27)

QUALITY AS VALUE

What is noteworthy about the criteria suggested by these media professionals is that they say almost nothing about audience size. Only the last of the questions above implies that ratings may, at some level, be a necessary — though not sufficient — indicator of quality.

Admirable though this vision might be, to minimize ratings in this way is of little help to public television programmers who struggle to survive in the real world. It may be the case that in other forms of artistic expression audience size is completely irrelevant — so that a brilliant work by a genius composer or novelist may go unrecognized for years, or even generations. But public television must produce credible results in real time, because it is paid for by the public. In the real world of politics and public accounts committees and auditors-general, the public expenditure needs to be justified by some concrete indication of *value*.

"Value" is another of those slippery words that needs to be pinned down and parsed. There are at least two relevant meanings here. One is what could be called socially instrumental value; the other, moral value, as in that which is innately valuable or worth valuing. A compelling war bonds poster or a catchy advertising jingle, or news coverage of an election, might fit the definition of instrumentally valuable; a Turner landscape or a Beethoven sonata or a TV drama like *Wallander* could be said to have innate moral or aesthetic value.

The purpose of public service broadcasting in any country is not just to satisfy the desires of individuals for information and entertainment, but also to foster social cohesion and to act as a civilizing influence. These are mostly instrumental values. At the same time, we expect public service programs to be of authentic moral, ethical, and aesthetic quality as well. The mandate is to produce value in both instrumental and normative senses. In assessing public broadcasting, a connection needs to be made between quality and value.

The Nordicity research group has recently done some analysis along these lines, as part of a larger study called "Analysis of Government Support for Public Broadcasting and Other Culture in Canada" (April 2011). In a survey of eighteen industrialized countries, the study rated the potential value of public service broadcasting across five areas of potential risk and benefit to any nation. These included: population density; whether and how well various linguistic groups are being served; the level of ethnic/linguistic tensions within the nation; the presence of neighbouring countries which are major cultural exporters and share the same language(s); and the popularity of domestically-produced programming as indicated by the number of such programs currently in the top ten viewership ratings.

The study ranked Canada at the top of the list of countries likely to benefit from — i.e., countries in need of — public service broadcasting. To put it another way, Canada ranked first as the country where public service broadcasting was likely to provide the most value. The US, Japan, and Italy shared the lowest ranking, as culturally cohesive, densely-populated countries served by a diversity of broadcasters in the dominant domestic language, and with little same-language competition from outside. The only countries that came even close to Canada's rating were New Zealand and Switzerland.

The study is fascinating and well worth reading in the original. Among many startling statistics: in almost all of the eighteen nations studied, domestic programming accounted for 10 of 10 or 9 of 10 of the top ten rating positions. In French-speaking Canada it is also 10 of 10, while in English Canada it is 1 of 10! (The other nine are American.) And, the average subsidy paid to public broadcasters by eighteen Western countries surveyed by Nordicity in 2009 is about $87 per capita: in Canada it's about $34. (Roughly nine cents a day.) Norwegians pay about $164 a year; Danes pay $142; in the UK it's about $111 per capita; in Australia $44.

The study suggests, in other words, that although Canada ranks far higher than other, comparable, countries in being likely to benefit from public service broadcasting, our investment in the area is well below average. And the trend is in the wrong direction. In fact, between 1991 and 2009 the federal government's expenditure on CBC/Radio-Canada rose by only eight percent, while federal spending on other forms of culture went up by 71 percent, and total federal spending, excluding national defence and debt interest payments, rose 83 percent.

We've already seen how advertising-supported television is the prisoner of a market dynamic that ordains mediocrity via Gresham's Law: the bad drives out the good. Advertising-free public broadcasting, because it escapes this ineluctable market dynamic, has the opportunity to rise above mediocrity, and even to achieve excellence on a routine basis. Assuming the requisite skills and talent are available (and in Canada they certainly are), the only thing that stands between public television producers and this kind of brilliance is money, for high-quality programming is expensive to produce. Money, and I should add, the kind of management that is willing to take risks – and knows quality when it sees it.

Evaluating quality in television programming is something production professionals do all the time, using criteria that are widely accepted across the industry, including those discussed here. These judgments are regularly reinforced by contests like the Emmys, the Edward R. Murrow Awards, the Peabody Awards, the Geminis, the Golden Globes, the BAFTAs and scores of others. Value, however, is territory less explored. The criteria used in the Nordicity study are a good starting point for tackling this chore. Establishing national consensus on value provided by CBC/Radio-Canada would be a giant step toward putting our public broadcaster's future on a firm footing. It can be done, and the way to begin is to ensure that everybody's speaking the same language when they talk about quality in television.

5

Public broadcasting and the news

In 2009, CBC television news completely revamped its on-air appearance as part of the new, populist, ratings-focused orientation of the network. The set for the flagship evening newscast, *The National*, went from a traditional desk and over-the-shoulder graphic arrangement to something borrowed from *Star Trek: The Next Generation*. Frank N. Magid Associates, an Iowa-based television consulting firm that had made its name decades earlier by inventing the commercially successful and widely adopted "Action News" format for local TV news, was enlisted to provide instruction in the American way to success-through-ratings. What was designed to be a stable roster of correspondents was selected and groomed to provide a more stylish and animated approach to their on-camera presentation (plenty of hand gestures; involvement in the story by, for example, wearing hip waders for flood coverage, etc.). What had previously been an occasional use of faked "throws" from reporter to anchor and vice-versa, to give the appearance of live interaction, was de rigueur in the new show.

In the new newscast, huge background graphics screens were in perpetual motion, with text and colours drifting across the viewers' field of vision while the anchor, inexplicably standing, or even strolling across the set, introduced succeeding items like a circus-tent ringmaster. All news programs are scripted; this one seemed to be choreographed as well.

The Toronto Star's TV critic, Greg Quill, commented that CBC-TV's "now-unseated anchor, Peter Mansbridge, plays a cross between a wandering, gracious maître d' and — when he's standing behind the new Plexiglas counter — an avuncular publican pressing messengers to unburden themselves." (October 27, 2009). Quill went on to identify the most troubling aspect of what had been billed as a change in the program's look and feel, but was really much more. "What doesn't sit well, however, is the surreal, theatrical nature of the new *National.* It's less a news program than a piece of rehearsed drama. Characters step up to their marks — including retired general Rick Hillier, who bellied up to Mansbridge's pristine bar to deliver a broadside about Ottawa's ineffectiveness in backing up Canadian Forces in Afghanistan — to speak a part." This was the newscast as reality TV, or vice-versa.

To completely renovate a network newscast in this way is a very expensive proposition and in 2009, the CBC was, as usual, coping with crippling financial uncertainties and revenue shortfalls. Why, then, spend a small fortune on cosmetics?

For Richard Stursberg, it was part of a badly overdue modernization of a news operation that had not kept pace with its flashier counterparts in commercial television, where CNN led the way with its holograms and outsized graphic displays and Hollywood soundstage-sized set and roving presenters. Without that modernization, he believed, the public broadcaster would continue to fall behind its competitors in ratings, and would eventually cease to be relevant to the taxpaying citizens who were its intended audience. In retrospect, it was money ill-spent, not because of its lack of impact on ratings, but for more complex and interesting reasons.

None of this complexity seems to have been appreciated by Stursberg who saw news as a simple, uniform commodity. In his memoir he states that, whereas Canadian TV drama was in short supply, this was not the case with news. "There was no shortage of Canadian news," he says, listing the national newspapers, websites, radio stations, and network television newscasts available everywhere. "In no sense is there a market failure requiring remedial government." CBC news personnel have argued that there is a need for a "deeper, more thoughtful" approach to the news than is available on commercial networks, he notes, but he goes on to point to the fact that CBC television news is not the ratings leader, and to ask, "why did the CBC have to offer a news service when there was no shortage of Canadian news? What was the point of spending public money financing a service that the private sector was already providing?" (168) I'll try to provide some answers to these questions in the following pages. But first...

A CAUTIONARY TALE

When I arrived at CBC television news during the brief presidency of Tony Manera (1993-1995) it was as a consultant to the Vice-President of television news, Tim Kotcheff. Manera had hired him away from CTV, where he had been head of the news division, and where I had spent more than a decade writing and producing and eventually overseeing policy and business development. Technology was changing at lightning speed from analog to digital, from terrestrial distribution to satellites and cable; CBC had a newly minted all-news cable service called Newsworld; budgets were, as usual, under the axe and the organization suffered from an understandable case of bunker mentality. The expectation was that Kotcheff, with his training in private broadcasting, would be able to shape up the news and current affairs divisions, make them accountable to management, and trim away some of the "waste" those outside the organization were always sure was scandalously present.

Journalists are at the best of times cantankerous when it comes to defending their turf from the administrative and financial sides of a news operation, and with good reason. Ultimately, the defence of truth and fairness in a journalistic enterprise rests with the individual journalist, who must be prepared to sacrifice his or her job on the altar of principle. This is the only reliable bulwark against manipulation and distortion imposed from above, since democratic governments have, with good reason, traditionally preferred to leave the news as free from regulation as possible. It is a lesson I learned in my first newspaper job when a crusty old copy editor shared his credo with me: "Always edit with your hat on."

Kotcheff was a journalist of the old school in this regard, and his focus was on securing the independence of news from the skittish senior managers in Ottawa, who seemed to regular workers like the gods of Olympus — venal, profligate, fatuous, and at the same time all-powerful. But he was used to the rigid lines of authority that prevailed in private industry, and he expected his orders to be obeyed. In the simple-minded management rhetoric of the time, he wanted to create a team, of which he would be the coach. And to that end he hired consultants to put news producers through the usual, excruciating, "team-building" exercises familiar to corporate workers everywhere at the time. To their credit, the producers resisted mightily.

There were other challenges. Ivan Fecan was head of English television at the time. (He would soon become CEO of CTV.) He sent a memo to Kotcheff one day asking why CBC news should not drop out of competition with the private networks and instead concentrate on the kind of in-depth reporting and analysis the privates avoided as being too expensive and too boring for audiences. Kotcheff was alarmed, and I was given the task of writing a response, which I dutifully did over the next few days, explaining why competition is so vital to a healthy ecology of journalism in any country. The CBC, with its public money, had a responsibility to support that system, and for it to pull back from the day-to-day competitive struggle would weaken not just us, but our private competitors as well. It was an impassioned argument against Fecan's proposal, and Kotcheff delivered it forcefully in a boardroom meeting. That was among the first of many nails in his coffin. And in retrospect, I can see that we were both far too enamoured of competitive rivalry in the industry, and not creative enough to see how something better could be built outside the commercial framework.

Kotcheff stirred up a pack of sleeping dogs closer to home when, in the name of efficiency, he insisted on amalgamating the staff and facilities of *The National* and those of *The Journal* (an interview and documentary showcase that followed the news), that had until that time been completely separate entities representing the long tradition in public broadcasting of a distinction between news and current affairs. This, too, was probably misguided, though at the time it seemed merely common sense.

In Kotcheff's two-year tenure we launched Newsworld International, a revenue-spinning direct-to-home satellite news service (sold off to US interests in 2000 for $155 million, by then-president Robert Rabinovitch), and found a way to reinstate local news in Windsor, Ontario, by collaborating with our unions in developing a business plan. We worked hard on strengthening our relationships with other news leaders such as ITN, BBC, CBS, CNN and Mexico's Televisa. We tried to steal the high-value ABC News affiliation away from CTV. We experimented with new, digital production equipment just appearing on the market. We squirreled away assets such as mobile production trucks under the budgetary protection of Newsworld. The idea, in short, was to be as entrepreneurial as we could be, without compromising the basic integrity of the news programming we were producing.

In the end, though, Kotcheff's outsider status did him in. When Tony Manera resigned over the government's failure to keep its word on stable CBC funding, Kotcheff was left vulnerable without his powerful sponsor and was ousted in a palace coup like a rejected organ transplant. That left me in limbo; I had done my job well, made some friends, and there was no cause for ending my contract prematurely. I was given an office in the bowels of the Broadcasting Centre and absolutely nothing to do. My phone never rang; no one visited me. I began writing a history of radio, which expanded into a history of communications technologies and their impact on society, published a year later as *Spirit of the Web: The Age of Information from Telegraph to Internet*. During that time the terrible genocide in Rwanda began to unfold. I had been in that country researching a book a few years earlier, and tried in vain from my lame-duck position to interest television news in giving the story the treatment I felt it deserved: the executive producer was on his annual leave, and no one else would risk making the investment required to organize coverage.

The impression I was left with when I finally packed up my books and papers and left the Broadcasting Centre for good was of a television news operation that had become its own worst enemy. Its defensiveness and insularity were no longer there to protect it from manipulation by vested interests, but to avoid the kind of upper management scrutiny that would have revealed the extent of nepotism and old-boy favouritism that had corrupted its senior echelons. Or so it seemed to me. Many of the very best, most dedicated public servants I'd met there either worked full-time for one of the unions, or would eventually jump ship to the rapidly expanding world of independent production, cable specialty channels, and Internet start-ups. The thinning of the ranks of the most talented and dedicated seemed to me to have been a perfectly natural and predictable outcome of the inconsistent and often incompetent leadership coming from Ottawa head office. And in retrospect it is clear that these senior managers were struggling under impossible financial expectations, caught between a demanding public service mandate and successive governments unwilling to pony up the money needed to adequately fulfill that directive. It was nevertheless very sad to see such an important institution in such pathetic disarray.

Today I can see that Kotcheff's and my own acceptance of the entrepreneurial premise – that television news could be made to support itself financially without compromising editorial standards and public service goals – was wrong and fated to fail. For the CBC, like any large bureaucratic structure that is being starved for the money it needs to do its job properly, new sources of revenue always tend to become a focus of corporate energies. The same pressures and distortions of purpose face the university where I now teach, and in fact universities everywhere. At the CBC this was to the detriment of mandated public service responsibilities: that is, producing quality newscasts for a national audience. In universities it is to the detriment of educating, rather than simply training, students. In both cases the higher goals, however difficult to articulate and quantify, are the irreducible essence of the mission.

I can also see that the circling-of-the-wagons instinct that made the news and current affairs operations so difficult to manage from the top down, as we had tried to do, was a predictable reaction of responsible journalists to a situation in which successive presidential and vice-presidential regimes — men (they were all men) plucked from bureaucratic posts with no specialized knowledge of media or news — had sought to shape the organization along corporate industrial lines.

Apart from management meddling (for better or worse) there was at CBC, as elsewhere in the news business, a real danger of editorial interference. One of the ways in which influence is exercised over news media by the powerful vested interests is through "flak." This is simply the harassment of journalists by representatives of those powerful interests, who may be lawyers, public relations companies, pseudo-grass roots organizations, politicians, or senior corporate executives who network with the senior executives of the offending journalistic enterprise. Thus, the only time in a long career in journalism that I experienced direct pressure to alter a story on behalf of a commercial sponsor was when I was senior producer at *Marketplace*, the CBC's long-running consumer affairs program. The program itself was commercial-free, but a company whose automobiles we had criticized was a big source of advertising revenue for the network. The flak we experienced over that item came not from our direct superiors, who did what they could to shield us, but from the highest levels within the corporation, managers who were not involved in production, and whose offices were in Ottawa.[19] At the CBC, as in other large journalistic enterprises, it is the senior managers who most frequently and directly experience flak and transmit it down the food chain. In this case, top officers of the offended corporation contacted CBC brass directly to air their grievance and demand redress.

The journalists whom I and others had found so difficult to manage from our carpeted corner offices in the CBC's Toronto Broadcasting Centre were editing with their hats on.

TV NEWS AS PUBLIC SERVICE

Television news matters. Despite encroachments from the Internet, it remains an important, often the most important, source of news for ninety percent of Canadians.[20] And we rank it far above Internet sources such as blogs and social media in terms of reliability.[21] We count on it.

I have already described the operations of Gresham's Law – the bad drives out the good – and its impact on commercial TV programming. This dynamic is of particular concern in terms of television's role in establishing and maintaining democratic values and the "marketplace of ideas," where it inevitably leads to a relentless pressure on costs associated with programming such as news and documentaries. It militates both against the allocation of air time devoted to maintaining public space in media and against the quality of content appearing in that space. Free market dynamics, far from serving the public interest in a lively public space, conspire against it. The evidence accumulated over three decades of deregulation in media markets strongly supports this conclusion: eviscerated newsrooms; abandoned foreign bureaus; the industry-wide move toward "infotainment;" continuing concentration of ownership; and a continuing decline in public trust, which can be directly related to the decline in content quality.[22]

To say that Gresham's Law is an "inevitable" consequence of commercial sponsorship goes against the grain of academic thinking in the social sciences these days, but I use the word advisedly. In a world in which major media outlets are owned by a handful of mammoth corporations, the notion that profit might be sacrificed for public service is literally absurd. Modern business corporations are supremely rational entities and require that every aspect of their operations conform to the logic of profit. [23]

But corporations are also creatures of the state, and the state can regulate their behaviour. And where news is concerned, state regulation of broadcasters was a feature of the market in most industrialized nations throughout much of the twentieth century, up until the rise of neo-liberal economic ideology in the 1970s and 1980s. In the United States, the swing to neo-liberalism and, with it, deregulation, roughly coincided with the purchase of the largest three broadcasters, ABC, NBC and CBS, by new corporate owners.[24] Deregulation meant relaxation or elimination of statutory obligations that had required broadcasters to maintain professional news operations and present newscasts of prescribed length at prescribed intervals during the broadcast day. The ownership changes placed the major networks in the hands of corporate entities whose MBA-trained senior managers regarded news divisions as potential profit centres and insisted that costs be reduced and revenues increased accordingly. The new entrepreneurial culture dictated that news programming be made more arresting and entertaining, so as to broaden its audience, making it more attractive to a wider array of advertisers.[25]

The information quotient of news was progressively sacrificed to entertainment in the form of celebrity news, soft features, spectacle, and sensation. The notion that one of the jobs of a responsible news organization was to present its audience with what it ought to know – even if it meant boring some of them some of the time, and even if it meant sometimes spending money on an investigative story that might not pan out – simply did not fit with the new business ethic. Since then, Gresham's Law has proved to be alive and well in commercially sponsored news operations across media platforms. The rationalization process that swept through the industry has bequeathed us, not merely dumbed-down newscasts across the board, but the aggressively mendacious Fox News in the US and its tiny Canadian clone, the Sun News Network, not to mention a toxic cauldron of American talk radio stations.

Arthur Kent, the Canadian journalist who gained fame at NBC News for his Gulf War reporting, has described the devolution process as he experienced it from the inside. He reports that the slide toward tabloid news began immediately after General Electric bought the network in 1986. Management of the news and entertainment divisions was consolidated, with a strict mandate to increase revenues. NBC News had operated at a loss from 1979 to 1988 — as was the accepted state of affairs in news divisions at all three major networks prior to the spate of takeovers. Staff were laid off, and long-standing policies against hiring outside investigators to work on stories, and paying for interviews, were overturned. The man GE hired to run the news division resigned in 1992 over a scandal involving the rigging of explosions in GM trucks in a *Dateline* story about faulty gas tanks. For Kent, this was symptomatic of the decline in professionalism the news division had suffered under GE's aggressively ratings-and-profit-oriented stewardship.[26]

Public confidence in television news has reflected this realignment of priorities over the past three decades. The Gallup organization's annual polling of the confidence Americans hold in various public institutions ranks TV news identically with "big business" and banks —just 21 percent of respondents saying they have "a great deal/quite a lot of confidence" in television news; 38 percent have "very little or none." This puts TV news ahead only of Health Maintenance Organizations (HMOs) at 19 percent and Congress at 13 percent. When Gallup started tracking confidence in American television news in 1993, confidence level was at 46 percent, and other evidence suggests it had been even higher a decade earlier.[27] By contrast, in Europe and Canada, where public broadcasting continues to play an important, often dominant, role in the media mix, confidence in television news remains conspicuously higher than in the US.[28]

As a hybrid, quasi-commercial operation, CBC television news must constantly try to balance its public service mandate with its role as a revenue centre for the network. This has led, as with the rest of the network's operations, to a certain schizophrenia that manifests itself in otherwise inexplicable policy decisions. These include (but are by no means limited to) changing the air-time for the flagship *The National* from 10 p.m. to 9 p.m. and back again, disbanding its crack documentary unit, preempting the newscast for NHL hockey playoffs, preempting the news for a disastrous reality program called *The One*, and the recent consultant-driven, look-and-feel renovation which has tried to mimic CNN. In its internal struggle with its conflicting demons, the commercial imperative is winning, and CBC television news has slowly but inexorably edged in the direction of mass marketing and entertainment: in the direction of news as performance. [29]

The personalizing of CBC News in the figure of the flagship television newscast's presenter, Peter Mansbridge (grandiloquently called "chief correspondent") is a method filched from corporate branding – think of Quaker Oats' Aunt Jemima, GM's Mr. Goodwrench, Proctor and Gamble's Mr. Clean. And as a technique it is expensive both in terms of salary and power relationships in the news production process. The anchor, willy-nilly and regardless of intellectual acumen or journalistic perceptiveness, becomes the 900-pound gorilla in the newsroom because he or she is the franchise, the immediately-identifiable corporate face of a lucrative enterprise.

The question of how, and to what degree, the public broadcaster ought to compete with commercial networks in news is an issue that needs to be addressed with some disciplined thinking, rather than on gut instinct. Any journalist will tell you that news is by its very nature a competitive business. Journalists are motivated by the "beat;" by being, as Edward R. Murrow's CBS News used to boast, "first with the best." It is one of the things that make the profession so exciting to work in. The *Washington Post's* Watergate exclusives led *The New York Times* to devote expanded resources to an energetic effort to match the competition, and the result was good for the public. A similar, contemporary, case could be made for the continuing coverage of the "robocall" voter-suppression scandal in this country, in which a variety of news outlets, including the CBC and Postmedia, challenged one another for the lead. This kind of competition is generally invisible to the casual observer, and its rewards are typically of the kind that only media professionals recognize and value. Nevertheless *competition within and between news organizations* to be "first with the best" is healthy and productive of public good, and should be encouraged. It is *competition for audiences*, as in ratings competition, that winds up being so corrosive to standards and values.

FINDING A BALANCE

In considering the role of news in public service broadcasting, it is important to consider the ecology of news. Within that ecosystem the public broadcaster has several responsibilities:

- to provide a benchmark for quality — that is, to compete strongly in conventional news coverage to encourage private broadcasters to be the best they can be;
- to provide the kind of depth and context that private broadcasters find it unprofitable to produce;
- to cover stories of high importance but low audience appeal, such as ceremonial events of national significance, political conventions, etc.;
- to provide a consistent, reliable, cumulative historical record for the nation, from all regions of the country and from significant foreign locations;
- to comment critically from time to time on the overall health of the journalistic ecosystem within which it operates;to provide a competent, independent, and reliable source of emergency regional or nation-wide communication when needed.

If it does these jobs well, the role of ratings in judging the success of the news operation fades in importance. The public broadcaster's responsibility is to the citizenry at large, and not to sponsors or corporate shareholders: audience numbers need to be considered in the broader context of the news available from all competing outlets within the system. If CTV, the current market leader in nighttime national newscasts, is providing excellent coverage of national and world news, then it deserves high ratings, and the nation is better off for their success. Presumably, Global will push hard to gain market share at CTV's expense, reaping increased ad revenue in the process. However, if CTV achieves its ratings lead by resorting to the kinds of counter-journalistic methods employed on Fox News, then the role of the public broadcaster becomes even more important in providing an industry benchmark for quality.

It is not difficult to imagine a first-class news operation whose programming does not do well in the comparative ratings game – the US market provides an obvious example with excellent but obscure NPR and PBS news services up against Fox News, until the debacle of its 2013 presidential election night coverage the cable news ratings leader. By the same token, the BBC provides ample evidence that excellence need not be equated with invisibility; in fact the BBC's leading position in its market and around the world indicates that genuine quality may well be rewarded with ratings success. The benchmark provided by such a market leader forces all of its competitors to strive for the same quality. Thus, ITN, which served Britain's commercial networks, initially provided comparable quality to the BBC, though in recent years it has fallen victim to ill-conceived attempts to inject more competition onto the business of providing news for the nation's commercial broadcasters.

WHAT IS NEWS?

Finally, it's worth looking briefly into the most basic question of all, because attempting to answer it sheds light on the fundamental differences between commercial and public service news services. What is news? Events themselves are not news: events are what they are – a train crash, the passage of a bill, a war declared. "News" is the concise, reliable report that tells people about an event. In other words, news is *a report* of an event, and a report is quite different from an event. It is an artifact, a construct, a symbolic representation. And in the world of commercial media, news becomes something more than that, as well: as a report, it becomes a product available to be bought and sold.

But clearly not everything that happens merits the preparation of such a report; not everything that happens is considered worthy of being packaged as news. A process of selection is involved. And following that, a process of production. Each of these processes introduces a great many variables and contingencies into the making of the final product as it is seen by audiences.

Selection is guided, first of all, by two interlocking variables: first, the interests of the audience, where "interests" is used in its widest sense of common concerns and curiosities as well as that which is seen to be of importance or significance; and second, the judgment of news professionals (reporters, editors, producers).[30] Because audiences have little input into these selection and production processes, in the real world it is mainly professional journalistic judgment that determines what gets made into the news commodity. This amounts to journalistic judgments as to what is of importance to the audience on the one hand, and what the audience is likely to be curious about, or amused by, or amazed by, on the other. These professional judgments as to what has meaning or significance, and what has entertainment value, are of course subject to the biases and prejudices of the individuals making them, which are in turn heavily influenced by prevailing political and economic beliefs. In this part of the world, for example, we all swim like fish in a sea of liberal capitalist, materialist, individualist, consumerist ideology, seldom stopping to consider that there are other ways of organizing society.

And to further complicate matters, keep in mind that "the interests of the audience" are to some degree determined by what the news media present to them. For one thing, audiences can't be concerned about what they don't know. Furthermore, what they "know" via the news media may or may not be truthful or complete or accurate, or it may have been ascribed a false significance.

Lastly, professional judgment as to what will constitute news on any given day is more deeply influenced than most news professionals would care to admit by such considerations as whether events are accessible and coverage is likely to be affordable. These choices are in turn contingent on the particular business model under which the news producer is operating: all things being equal, in a for-profit, advertising-supported model, greater weight is likely to be given to cost in the cost v. public benefit equation than it is in the public service model.

In all of these areas of determination, the fact that public broadcasters are oriented exclusively toward serving the public, while commercial broadcasters are torn in an unequal battle between public service and profit, makes a tangible difference in what gets covered as news, and how it is presented. Commercial outlets will, for example, tilt news production away from longer documentary-style items (expensive to produce) and toward commentary by individuals or panels (generally less informative, but livelier and cheaper to produce). It will dictate shorter, rather than longer newscasts, and shorter items within those newscasts. It will focus editorial attention on viscerally appealing subjects such as crime, disaster, tragedy, and extreme weather, even though they may have little real significance for the audience. Over time, it will deteriorate into wall-to-wall infotainment.

The need to tailor the news to the constraints imposed by advertising — both the time constraints and the need to entertain as wide an audience as possible — have led over the years to a virtually homogenous style of news presentation on television. It involves dramatic story-telling structures; an emphasis on conflict rather than consensus; an aversion to "process" stories in covering government; an avoidance of complexity, especially in numbers; the use of telegenic authority figures and experts; an aversion to "talking heads;" a tendency to reduce all stories to two opposing sides; the use of celebrity stories and other "fluff" as a leavening agent; the use of reporters as participants in their stories; a preference for strong visuals wherever possible, and so on.

The veteran NBC news producer Reuven Frank pointed out many years ago that television news operates at a disadvantage to the newspaper when it comes to providing a balanced menu of information to its audience:

> A newspaper...can easily afford to print an item of conceivable interest to only a small percentage of its readers. A television news program must be put together with the assumption that each item will be of some interest to everyone that watches. Every time a newspaper includes a feature which will attract a specialized group, it can assume it is adding at least a bit to its circulation. To the degree a television news program includes an item of this sort...it must assume its audience will diminish.[31]

Apart from their central insight, Frank's comments are interesting in that he made them in the early 1970s. This was what might be considered the golden era of American television news, prior to the takeover and hyper-commercialization of the American networks, after which the focus on ratings and profitability for news was much intensified. Even at that time, however, it was clear that there was a need for an alternative to the commercial product in a world where most people get most of their news from television. That need is greater today than ever.

6

A few words about radio

In 2002, a 112-car Canadian Pacific freight train derailed just outside Minot, North Dakota, pop. 42,000, the state's fourth largest city. It was just before 2 a.m. The train's conductor made a 911 call to the city's emergency dispatcher warning that cars carrying hazardous materials appeared to be leaking. There was a pungent, suffocating odour in the air. It was quickly determined that the hazardous material was anhydrous ammonia; 240,000 gallons eventually leaked and vapourized, sending a toxic gas plume drifting across the city. Even low-level contact burns the eyes, nose, and throat, and exposure to high levels can cause death from a constricted throat or from chemical burns to the lungs.

The local 911 system was deluged with calls, and operators told panicking residents to tune into the designated emergency radio station for information and advice. But when residents tuned in, all they heard was canned music – normal programming. When they telephoned the station no one answered. Other stations were no better.

There were six radio stations in Minot, and throughout the small hours of the morning, all of them continued broadcasting music and chat as if nothing was happening. All six stations were owned by Clear Channel Communications, part of a hugely profitable stable of about 850 radio stations held by the company across the US.[32] The stations are highly automated, and programming is syndicated, produced, and packaged many miles away from Minot. That night, one man died and hundreds of other people were hospitalized. Thousands spent a terrifying night not knowing what was going on.

The Minot story is an extreme example of what has been happening everywhere in the US since the elimination in 1996 of ownership regulations that had limited any one company from owning more than one AM and one FM station in any given market. Every day, people desperate for information on approaching tornadoes or flash floods or blizzards or road closures or power failures or other emergencies turn to their radios for information, only to find that automated programming from some remote location is all they can find.

Radio has always been an essential component of any national communication system. In emergencies, people turn to radio for information, because radio is local, portable, and can be accessed on inexpensive battery-powered appliances even when power grids are down. In a deregulated world where mega-corporations like Clear Channel can own unlimited numbers of stations in any given market, commercial radio no longer fulfils even this most basic of public service functions, simply because it is not cost-efficient to staff stations with real people when they can more cheaply be run by computers.

The situation in Canada is not quite so dire as it is in the US, but the continuing trend to corporate mergers in the industry is putting ownership of radio stations in fewer and fewer hands, and the mood in government in Ottawa continues to be one that favours less and less regulation. For the moment, however, there remains some diversity on the Canadian airwaves, and we are preserved from playlists identical to those on American stations thanks mainly to the CRTC's Canadian content regulations. Can-con rules began in the 1970s with a mandatory 25 percent Canadian content rule, which has since risen progressively to 40 percent in new licensing. Although commercial broadcasters initially complained loudly that Can-con regulations would doom their industry by driving away listeners, the opposite has happened. It is by now an accepted fact of Canadian cultural history that these regulations led to the development of a vibrant Canadian music industry where none had existed before, and commercial radio continues to thrive as an adjunct to that industry.

In 1975 CBC radio eliminated commercial sponsorship and thus became, for the first time, an authentic public service broadcaster. This extraordinary policy shift was proposed by CBC President Laurent Picard and implemented during the tenure of his successor, A. W. (Al) Johnson. A career public servant of exceptional talent, Johnson held a Ph.D. in public administration from Harvard and had spent a decade as deputy provincial treasurer under Premier Tommy Douglas in the CCF-NDP government of Saskatchewan. He then moved to Ottawa to become assistant Deputy Minister of Finance under Prime Minister Lester Pearson, and he played an important role there in helping to develop Canadian's national Medicare system. Prime Minister Pierre Trudeau appointed him to the CBC post in 1975.

Johnson's left-leaning public service instincts combined fortuitously with the reality of declining advertising revenue at CBC radio (due largely to competition from the burgeoning television service) to allow him to take what seems today to be an impossibly radical step. At the time of the change, advertising was providing just sixteen percent of the radio service's total budget of $45 million. Johnson was also helped along politically by the commercial radio lobby in Ottawa, which had been strident in its insistence that the public broadcaster was unfairly competing with commercial broadcasters for advertising revenue, reducing their profit margins.

In the years following the removal of advertising, CBC radio experienced a renaissance in programming that increased the already intense loyalty of an expanding audience and led it to leadership in its largest urban markets. The absence of advertising, serious attention to the arts and the world of ideas, top-quality, thoughtfully-produced current affairs programming, and virtually 100 percent Canadian content proved to be a winning formula; the contrast with CBC television could not have been greater. At the helm of radio during those years was a remarkable, BBC-trained producer named Margaret Lyons, who transformed the broadcast schedule with new and innovative programs such as *As It Happens*, *Quirks and Quarks*, *Sunday Edition*, and *This Country in the Morning*. Perhaps more significantly, she set about making the public broadcaster sound more like the people it served — more colloquial, more conversational, less stilted and officious. To that end she cultivated on-air talents that have since become legendary: Barbara Frum, Peter Gzowski, David Suzuki, Michael Enright, Vicki Gabereau and many others.

A decade later, local and regional programming underwent a radical revision of its own, designed to ensure that programs in the morning and evening drive, and lunch time slots more closely reflected the diversity of the urban populations they served (Savage, 281-84). Serious and sustained effort was put into recruiting and training both on-air presenters and behind-the-scenes production staff from diverse ethnic and cultural backgrounds. These efforts were rewarded with solid ratings successes across the country. And the local success brought more and more listeners to the national programming: daily audience share for CBC Radio One in markets where there is a local station are uniformly double what they are in markets without a local presence. This experience could well be a model for revival of CBC's local and regional television. Quality can build audiences.

CBC radio is unique in its market: it has no competitors. Nothing on commercial radio bears comparison, either in terms of Canadian content, or in terms of intelligence and quality. CBC radio currently accounts for about 20 percent of radio listening hours in Canada, an impressive figure given the number of options.[33] Audience loyalty is legendary: when a CBC listener turns on her radio and doesn't like what she hears, she doesn't go to another station; she turns her radio off. Sadly, it is an open question whether CBC radio's reach will hold up now that the service has reduced original current affairs programming, eliminated drama, and resorted to much more frequent repeats of its programming in the face of yet further budget cuts imposed by Ottawa.

THE TRANSFORMATION AT RADIO 2

CBC launched a small, experimental FM network in 1948, on which it simulcast programming from the main AM network. Distinctive music and arts-oriented FM programming began in 1960, and in 1975 the network was renamed CBC Stereo. By 1997 virtually all of the original AM network had migrated to FM as well, and the branding identities were changed to Radio One and Radio Two. Until 2008 most of the music heard on Radio Two was classical and jazz, and like Radio One, the audience (significantly older than Radio One's, on average) was fanatically loyal, though more modest in number. In that year, another Stursberg-backed initiative transformed Radio Two (now officially Radio 2) into what in commercial radio is known as an adult contemporary format, dramatically reducing classical music content on weekdays to a little over four hours of predictable favourites during the mid-morning. It was an overt attempt to boost the service's ratings by tapping into a younger demographic. Stursberg had argued, "Radio 2's audience was aging.... If the senior service was not careful, it would lose the listenership of Radio 2 not to competitors, but to the grave." (233) Never mind that the "aging" demographic is undergoing a boom that will see it double in size as the entire baby boom generation turns sixty-five in coming decades.[34] Stursberg's candid view was that, in any case, the classical music on Radio 2, "had nothing to do with Canadian culture" (236). Listeners who had been accustomed to waking up to and driving home to symphonic music and jazz were now treated to the likes of Blue Rodeo, The Cowboy Junkies, The Tragically Hip, and Arcade Fire. Fine music, but not their cup of tea, and anyway it was music they could hear elsewhere on the dial. At about the same time, the CBC announced it was shutting down the venerable CBC Symphony Orchestra in Vancouver. There had once been five CBC orchestras; the others, in Winnipeg, Toronto, Montreal, Halifax, were disbanded after earlier budget cuts in the 1980s. The Vancouver orchestra had for sixty-five years played an important role in Canadian classical music by commissioning, performing and recording new Canadian compositions. It had been the last of North America's radio symphonies; commercial networks had long since concluded that there was no profit to be made in maintaining such ensembles as Arturo Toscanini's NBC Radio Orchestra. In contrast, public broadcasters in Europe currently maintain a total of more than fifty radio orchestras: the BBC alone has five. For Radio 2 aficionados, the changes were apocalyptic, and audience share collapsed. Columnist Jeffrey Simpson spoke for many when he wrote in *The Globe and Mail* that Radio 2 had been a "redoubtable" island of intelligence in an otherwise largely mindless broadcast environment. "Radio 2's distinguishing characteristic was its intelligence," he said. "It emphasized classical music because that kind of music was not easily available in private radio and because, throughout the ages, that form of music appealed to the intelligence and the deepest emotions of listeners...[it is] the deepest exploration of the human dilemma through music." (Mar. 29, 2008) A House of Commons Committee chastised CBC managers for their folly, members of all parties expressing their dismay. The corporation defended the change by citing a need to attract a younger audience in order to remain "relevant," and by pointing out that there was a world of new Canadian contemporary rock music that was not getting air play on commercial stations. (Radio

2, however, had featured it in successive late-night and weekend programs including *Brave New Waves, Nightlines,* and *RadioSonic,* and the new, online Radio 3 carried nothing but.) The new Radio 2 format was designed to give exposure to Canadian groups that were less likely to be played on commercial rock stations, where mandatory Can-con time was devoted to well-established names, groups whose music you could take to the bank. Thus, Radio 2 listeners, long accustomed to the likes of Copeland, Rachmaninoff, Smetana, Górecki and Pärt at dinner time, were now treated to journeyman rock bands, "emerging" singer-songwriters, and their shrieking (as opposed to applauding) audiences on the new *Canada Live* concert series. The producers seem to have missed the point that most rock concerts, even those performed by legendary bands, are much more about the experience of being there than the music itself, which is often, to be kind, rough around the edges when compared to studio-recorded renditions. If there is a program genre that is not well-suited to radio, this is probably it.

To mollify listeners, CBC introduced at the same time a clutch of audio channels on the corporate website, which played continuous streams of classical, jazz, singer-songwriter and contemporary Canadian classical compositions. Stursberg explained the new service this way: "'You want classical music?' we would ask. 'Just go to the CBC website and stream it on your sound system. There it will be. No commercials. No annoying commentary. Just classical music twenty-four hours a day.'" Those who felt this was a lame substitute for the intelligent presentation that had been the hallmark of Radio 2, Stursberg ridiculed as dimwitted fogeys (238-39). Time will tell whether the steadily-expanding offerings of CBC's online music offerings at cbc.ca/music will prove popular. But it should never be forgotten that without the intelligent commentary of informed presenters, these services forego the informative, educational function that defines a public broadcaster. Providing supporting text at the website helps, but isn't a replacement for a live host. Nor is an atomized collection of solitary downloaders a replacement for an actual audience. Streaming audio (or video) fulfils only a small part of the public broadcasting mandate.

There is some evidence that audience numbers for Radio 2 are slowly beginning to claw their way out of the abyss into which they had fallen, and on-air personalities such as Rich Terfry, Tom Powers, and Laurie Brown are developing devoted followings comparable to those of earlier classical music hosts like Eric Friesen and Jurgen Goth. Continual tweaking by talented producers has helped. It remains to be seen whether a menu of classical war-horses surrounded by mainly Canadian adult alternative rock can satisfy audiences in the same way that programming dedicated exclusively to either format might.

In the meanwhile, it is important to understand that behind the remake of Radio 2 lies the same tacit, erroneous assumption that was made by Stursberg about television: that is, that the medium imposes strict constraints on the content. In the case of television, he insisted that only light, narrative entertainment and live events like sports were suitable and appropriate to the medium. He was openly scornful of cultural programming like the primetime *Opening Night* performing arts series, shows that did not conform to the dictates of Hollywood-style dramatic, reality, and comedy formats. The only evidence adduced for this assumption about the limits of the medium was that this was the format that dominated American television (assumed to be a "success") and the fact that audience numbers were often lower for performing arts programs than for, say, hockey. But commercial television programming is what it has become not because of any parameters imposed by the technology per se, but because that is where profitability lay. It is the demands of advertising that shape television.

In the case of radio, the assumption appears to have been that since private radio has become almost exclusively a purveyor of pop music in various genres and niches along with a smattering of talk-and-news stations, this is the "appropriate" role for radio as a technology. Hence, for example, the tragic decisions to eliminate the radio drama department and the CBC Radio Orchestra. Again, though, what has actually determined the nature of content on private radio has been a calculus showing that what earns the most money is lowering costs and reliably serving up the kinds of audiences advertisers want. There is absolutely no reason in logic for the assumption that radio is no longer suitable for programming beyond the boundaries of pop music and news/talk formats. In fact, there is an excellent argument to be made for the continued exploration of the promise of radio and its potential new role in a world of broadband multimedia, podcasting, and streaming audio on all manner of portable digital devices.

A glance back at the history of CBC radio provides some indication of the possibilities (though, of course, not a prescription for today). While music, mainly live and orchestral, was a mainstay of early radio – it filled time at reasonable cost and kept audiences entertained – between 1936 and 1941 the portion of the CBC broadcast schedule occupied by music fell from 70 percent to 51 percent. The CBC drama department had been formed in 1938, and during its first year producers read through more than a thousand submissions from aspiring playwrights and broadcast 350 of them. Beginning with World War II, news and current affairs began taking over a larger portion of the broadcast day, but drama continued to be a mainstay. In 1941 the network produced a series of classic plays on themes related to the war against Fascism, starring some of the leading international stage actors of the time: Shaw's *St. Joan*; Ibsen's *An Enemy of the People*; Galsworthy's *Strife*; Drinkwater's *Abraham Lincoln*; Obler's *This Precious Freedom*, Shelley's *Hellas* and others. In the 1947-48 season CBC radio presented 300 plays, almost all of them works by Canadian playwrights. This does not include offerings from the flagship dramatic series *CBC*

Wednesday Night, which broadcast classics from Shakespeare, Marlowe, Tolstoy, Shaw, O'Casey, and Eliot. These works were produced in studios in Vancouver, Halifax, Winnipeg, Montreal, and Toronto using Canadian producers and technicians and, for the most part, Canadian actors. As broadcast historian Austin Weir reported: "Canadian talent was being given genuine encouragement on an increasing scale, and — most important of all — there was a growing spirit of creative work being accomplished, which made the Corporation increasingly attractive to young and talented intellectuals." (273)

In addition, CBC radio carried commercially-sponsored Canadian variety, music, and drama, and a much larger offering of American commercial programming, including *Edgar Bergen and Charlie McCarthy, Our Miss Brooks, Fibber McGee and Molly, Ozzie and Harriet, Kraft Music Hall, The Aldrich Family,* and shows staring Bing Crosby, Jack Benny and Bob Hope.[35]

As a model for serving special interests of all kinds, the CBC farm broadcasts of the 1930s, 1940s, and 1950s were exemplary. They engaged not only rural Canada but urban listeners as well, with their reality-based dramas about life among farm families. The model was later exported to developing countries around the world. Children's programming included iconic shows like *Maggie Muggins* and *Just Mary*.

The story of public service, creativity, and innovation on CBC radio could be extended for many pages. The important point is that radio as a medium presents endless possibilities for audience engagement, entertainment, and education, and it does so at about a tenth of the cost of similar productions on television. Does television do it better? Sometimes, undoubtedly, but there is good reason to believe that children's programming, for instance, is better on radio, where the child's imagination is fully engaged, and where the kind of stultifying, hypnotizing effect produced by standard children's fare on television is not an issue. For the same reasons, drama often works especially well on radio. Many sports fans prefer to listen to baseball on radio. Interviews of all kinds are frequently more intimate and revealing on radio than on television, just as comedy is often more creative and inventive. Why should it be assumed that well produced and acted, well publicized radio drama is obsolete as a legitimate vehicle of public service on radio? One of the last dramatic series produced by CBC radio was *Afghanada*, a highly engaging and often moving series following the lives of Canadian soldiers fighting in Kandahar province. It ran for more than one hundred episodes over five years and drew an average audience ranging from 300,000 to 600,000 a week on radio and online. By any standard related to public service, this has to be considered a success. Current programs like *Wiretap*, *Spark*, *This Is That*, *The Debaters*, *DNTO*, *The Current*, and *Q*, along with perennial audience favourites like *Sunday Morning*, *Ideas*, and *As It Happens* demonstrate that there is still plenty of scope for experimentation on radio. In our era of exploding technical possibilities in communication media, the CBC ought to be more fully engaged with radio than ever.

Which raises another issue. What many Canadians, CBC fans and foes alike, fail to consider is that serving the entire range of listener preferences, from pop and rock to classical, to jazz and world music to talks, documentaries and drama, comedy and variety, news and special events, on two over-the-air radio channels (in each official language), is simply impossible. In other words, it is not possible for CBC radio to fulfill the demands of its public service mandate with just Radio One and Radio 2 and their French-language equivalents. In Britain, BBC radio operates: Radio 1 (pop/rock/urban); Radio 2 (adult pop/alternative); Radio 3 (classical, jazz, arts and drama); Radio 4 (the arts, ideas, current affairs); Radio 4 Extra (digital service/archival selections); and Radio Five (live news and sports). Plus, the associated individual websites which support the over-the-air services, and which feature streaming audio, archives, and podcasts. There is no reason in principle why CBC Radio could not be similarly expanded to serve a wider audience on more frequencies, should Canadians demand it.

Just imagine.

Unfortunately, under current CBC management, radio budgets are being slashed to support the television services. A Canadian Media Research Inc. report based on CRTC filings for 2009-10 shows that while all of the CBC television services showed increases in expenditures over the year, CBC radio's budget was chopped by 9 percent or $35 million. The report suggests that most of that money went to the corporation's advertising sales department, where budgets soared by $20 million in the same period. A more recent CRTC filing shows that in 2011 radio suffered an additional $27 million cut, while television services again reported increased expenditures across the board.[36] The results are evident on your radio, with reductions in local noon-hour programming, the winding up of the legendary radio drama department, and much more frequent program repeats in the afternoon and evening. And, if CBC management can get the CRTC to agree, advertising will reappear on radio services for the first time since 1975, a giant step backward.

Here is further evidence of the corrupting influence of advertising on public broadcasting: the most popular, most admired, and only true public service operation run by the CBC is being systematically and scandalously starved in order to support the bureaucracy that sells advertisements for CBC television.

7

Public broadcasting and "new media"

There is a tendency, especially among the young "digital natives" in our midst, to think of all media as being fundamentally alike. After all, communication media, digital or analog, over-the-air or online, web-based or broadcast, are all designed to disseminate information of one kind or another, whether it be news, or entertainment, or educational material, or art, or advertising and other propaganda, or the much more personal communication that takes place in web environments like Facebook.

But there is an important distinction to be made between broadcast media and the interactive, two-way media typified by web services. Broadcasting is a one-to-many undertaking, in which the content-provision function is located at the peak of a pyramid, with many more or less passive receivers spread out around the transmitter's footprint. The web can serve this function as well, as it does with streaming audio or video, or on sites like YouTube, but it is genuinely revolutionary in its ability to allow interaction between and among content providers and users.[37] In fact the two are often the same person. In the interactive environment fostered by many web applications, there is no "audience" in the broadcast sense, and every user is or can be, a content provider. That is the genius of social media as a business model: users supply virtually all the content at no cost, and the owners, who provide the infrastructure, reap the rewards from advertising and, increasingly, from the sale of data about the users.

In the early days of the web, futurists wrote enthusiastically about the empowering potential for anyone and everyone with access to a computer to become a publisher with a worldwide audience. There is truth in this vision, as the phenomenon of the viral video or monster Retweet can attest. But we've discovered over the past decade or two that peoples' attention is a scarce commodity. In a world in which, at this writing, there are an estimated 2.1 billion active Internet users, a billion Facebook users (of a total of 2.4 billion social networking accounts), 110 million tumblr and WordPress blogs, and 120 million active Twitter accounts, it's not easy to be heard. Unless you're already famous: Lady Gaga had more than 18 million Twitter followers in late 2012.

Within the general frame of reference of broadcast media, the Internet can serve the same purpose as an over-the-air transmitter, moving programming from a production or storage centre to the audience (normally an individual, as opposed to a group). The difference is that, due to the very large storage capacity and the near-instantaneous accessibility in the digital environment, it is possible to support "on-demand" services via the Internet. Programming in virtually unlimited quantity can be warehoused on digital storage media so that audience members can retrieve it at their convenience, rather than having to wait for it to appear on a traditional broadcaster's scheduled transmissions. Where news is concerned, the web makes it possible for individuals to create personalized "newscasts," selecting stories from a menu of interest categories such as science, technology, politics, international affairs, and so on.

The on-demand feature of web-based media services is convenient, but it differs from authentic broadcasting in an important way. The veteran Canadian television producer Richard Nielsen has described the distinction eloquently:

> Unlike the Internet, broadcasting is not about supplying a library to which the public has access. Broadcasting assembles a congregation. It is comparable to a concert hall or other meeting halls in our larger cities. Broadcasting is designed to provide a communal experience, an experience that helps build consensus by its very nature, a consensus that should impose the disciplines on [production] talent that ensure that its standards will be high enough to serve that function. (9)

This suggests that web-based services, for all their convenience, can never entirely replace public service broadcasting as a tool for developing and enriching social solidarity. Personalized webnews services are in no way a complete substitute for professionally-produced national, regional, and local newscasts that keep listeners and viewers informed on a broad range of topics that are, or ought to, be of interest to them as citizens of the nation and of the wider world. Broadcasting, with its power to vastly extend public space and facilitate dialogue, has well-proven potential to shape society. And that is why, as Nielsen says, only "people who believe that such collective experiences foster creativity and advance human possibilities are willing to encourage it. Those who have other reasons for mobilizing public opinion will see it as a danger and attempt to minimize its influence or, as in the US, channel it to achieve specific ends dedicated to religious or political conversion rather than exploration." (15) Among the most messianic proponents of the web's potential for social change are, no surprise, many staunch libertarians.

All of this must be kept in mind when thinking about the CBC's online or new media role. Certainly, whatever the corporation does on the web, it should maintain its traditional broadcast capacity for the foreseeable future. At this stage of technological evolution, nothing can replace traditional broadcasting's public service role. Streaming classical music twenty-four hours a day with no introductions or comment by informed presenters in no way replaces what was being done by Radio 2 before the format conversion of 2008. The same can be said about any other musical genre: streaming content in this way ignores the CBC's mandate to educate and inform, and not just entertain, its audience.

A SUPPLEMENTAL SERVICE

But the web can be used in many ways to supplement and enhance the CBC's traditional services. To begin with, the ability of smart phones to access data wirelessly has meant that for the first time, television can be truly portable in the same way that radio, with its smaller, simpler receivers, has always been. Watching your favourite drama or sports event on a small smartphone screen may not be ideal, but it beats missing it. And you can do it virtually anywhere in the world. Web-based news can be revised and updated continuously, and cbc.ca news service acts as one of a handful of reliable and impartial sources of brand-name news in the vast morass of unreliable blog-based speculation and commentary.

Of course, the web enables audience interaction of many kinds, including, for example, voting during elimination reality programs like *Dragon's Den* or *Battle of the Blades*. It is assumed that these kinds of interactive features strengthen audience involvement with programming, but it will be some time before audience interactivity reaches a state of maturity beyond its current role as add-on gimmick.

The web also provides an environment for experimenting with new ways to package and distribute information, and with new forms of entertainment. Web-based programming of any genre can be produced and distributed at very little cost, incorporating audience interaction features impossible in traditional broadcasting. For these reasons, the web is increasingly being used to test-market television pilots.

The immense storage capacity and instant, random accessibility of digital media means that audio and video programming of all kinds can be archived and made freely accessible, as can important texts and documents. All of this means that the CBC, in cooperation with other public institutions such as archives, libraries, art galleries, and museums, can aspire to be not only a public broadcaster, but the nation's memory bank; a cumulative, continually expanding archive of Canadian culture, freely available everywhere, on demand, at minimal cost. This is an enormously valuable — and just plain enormous — undertaking, and it requires substantial capital and curatorial resources: a lot more money, people and equipment than is currently available to it at the CBC.[38] It is not a role that any profit-seeking enterprise could be expected to take on, or to adequately fulfill, but it is a natural role for a public service institution like the CBC, mandated to inform, educate, and entertain its public with "all that is best in every department of human knowledge, endeavour, and achievement."

As it is in conventional broadcasting, the CBC's role on the web at cbc.ca is one of providing service to the public rather than to advertisers. That service ought to be distinguished by its quality, and that means above all intelligence, reliability, and responsibility. The web, with its burgeoning blogosphere, is a Wild West of worthwhile content, conspiracy theories, salacious gossip, propaganda, character assassination, vituperation, and misinformation of all kinds. What passes for news is often nothing more than endlessly regurgitated rumour and unsubstantiated "fact" seeded by public relations consultants. So-called "iterative journalism" in which half-baked stories are published on blogs and then endlessly commented upon and "refined" by readers is the opposite of a responsible journalism that strives, through research, direct observation, and dogged investigation, to produce a body of reliable fact. It is clear that the impact of advertising on web content of all kinds, but especially blogs, is even more pronounced than it is in conventional broadcast media. The blogosphere is driven, for the most part, by advertising, and is if anything more obsessed with ratings ("hits" or page-views) than conventional radio and television. Its production model — indeed, its entire economic structure — is built on the premise that popularity equals success, and success testifies to quality. Most of the ads on blogs pay the blogger according to the numbers of page views, and search engines are configured so as to favour keywords in current circulation. Bloggers, whether self-employed or working for sites like Gawker.com or Huffington Post, are thus advised to focus on currently popular topics and to construe their headlines and text in the most controversial, sensational way they can.[39]

CBC currently sees its digital offerings as a potentially important source of advertising revenue, and is actively engaged in finding ways to please sponsors with innovative kinds of product placement and embedding.[40] If the CBC is to continue along this path, it needs to develop a strict and transparent code governing how and when advertising will be allowed to intrude in the public broadcaster's online offerings. At a minimum, advertising should not be permitted to interrupt content or to interfere with access, as it does when users are forced to view commercials prior to a program or archive clip. The job of a public broadcaster on the web is twofold: first, to provide citizens, its constituency, with an island of taste, responsibility, and intelligence in this widening sea of "truthiness," prevarication, and provocation. Secondly, its job is to identify and develop those aspects of web-based media that are most capable of providing a useful service, as this technology, still in its infancy, continues its chaotic and undisciplined development.

It needs to be remembered that historically, successive new technologies of communication do not replace earlier technologies, but supplement them. Radio did not replace the telephone; black-and-white television did not replace radio; colour television did not replace movies. Digital technology provides new appliances for the creation, storage and retrieval of all kinds of media content, but it does not replace earlier platforms or the ways in which audiences relate to them and use them. There will, for the foreseeable future, be both a need and a demand for traditional media, and the public service vocation of the CBC and other public broadcasters will be more valuable, and more necessary, than ever. Television viewing in Canada has remained steady at just under twenty hours a week over the past decade.[41] Internet use has climbed from 5.6 hours in 2004 to 10.5 hours in 2011, but it seems to be leveling off there.[42] Canadians spend less than half an hour a week gleaning news and information on the web. About half of those surveyed in recent polls still say they get most of their news from TV, and although that is down from 60 percent a decade ago, the decline is nothing like that experienced by newspapers: respondents who say they get most of their news from newspapers has nosedived from 21 percent at the beginning of the decade to 10 percent today.[43]

The CBC and private broadcasters will both have to adjust to the unfolding reality of the migration of television program delivery to the Internet, via such enterprises as Netflix, Google/YouTube, and AppleTV. These services operate outside the purview of the Broadcasting Act and effectively bypass the entire Canadian broadcasting regulatory framework. The trend seems to be toward providing an ever-expanding library of not just movies, but regular television fare such as first-run dramatic and sitcom series. In the industry, this is called "over-the-top" or OTT delivery. For commercial television networks in Canada, it is a threat to their basic business model of buying American programs at a steep discount and delivering them to Canadian audiences complete with Canadian commercials. It is worth noting, for example, that Netflix's market capitalization of $12.5 billion is seven times the combined worth of all Canadian commercial broadcasting entities, which means it has plenty of resources to outbid Canadian networks on any given property and make it available at below-cable cost to Canadian OTT subscribers. It is difficult to see how the CRTC can regulate this to protect the financial interests of commercial television broadcasters (Miller). In a worst-case scenario, it is possible to see private Canadian broadcasters succumbing to competition from OTT providers, quitting the conventional broadcasting business and merging with the enemy. In doing that, major vertically integrated conglomerates like Rogers and Bell could well increase their profitability, since they would no longer be obliged to conform to costly CRTC-imposed regulations on Canadian content.

Since the CBC broadcast schedule is dominated by Canadian content, it is largely immune to this threat. As long as it retains the appropriate rights to commissioned programs, it can either prevent them from being distributed OTT, or it can charge a fee for their OTT distribution, whichever best serves the public interest in universal accessibility. Revitalizing its own production studios to produce more proprietary programming not subject to contractual stipulations of private production houses would add an additional layer of safety.

Whatever happens in this swiftly evolving world of OTT distribution, it is clear that the Canadian public interest lies in preserving and strengthening the public broadcaster, our only reliable source of authentic Canadian content.

8

Revitalizing the CBC

In an October 2010 interview, Robert Rabinovitch, who was President of the CBC between 1999 and 2007, said the following: "I'm concerned about the future of the CBC, deeply concerned. I think what happened this past week (Bell Canada's purchase of CTV) is a game changer, and I really doubt that the CBC is going to be able to compete in the future. I think they're going to lose all of their sports properties." *Hockey Night in Canada*, he said, is "toast."[44] And when CBC television loses *HNIC*, he might have added, it loses about $100 million in advertising revenue, and is left with a 350-hour hole to fill with Canadian content.

Given its already perilous financial situation, the CBC as we know it could very well be "toast" as well.

Rabinovitch continued: "So you have to ask yourself what is the CBC for? I don't know the answer, but I do know the way it is right now is embarrassing. It's going to get worse and worse. Or maybe ... it should just stick with radio. Radio One, I think, is superb."[45]

The comments illustrate both the dire nature of the CBC's current predicament and Rabinovitch's fundamental unsuitability for the position he held when he hired that other stranger to the subtler aspects of public service media, Richard Stursberg. "What is the CBC for? I don't know." It could be their joint epitaph.

In the same interview, Rabinovitch called for a national discussion of public broadcasting and its future: "What I think we need in this country is a real debate... given the fundamental change in the environment that's occurred, where broadcasters are part of other empires, transmission [i.e. cable and satellite distribution] empires, with broadcast and music commodities."

Aside from the occasional blog troll who demands the immediate dissolution or privatization of the CBC, few Canadians would disagree with the need for public debate. The cornerstone legislation governing media in Canada, the Broadcasting Act of 1991, is approaching a quarter-century old — an eternity in the world of communication technology. Since it was passed, the media landscape has changed radically, and few of those changes have been of benefit to the public broadcaster as currently configured. The Act's assumption that public service is to be the guiding principle of *all* broadcasting, and its requirement that *all* broadcasting be "predominantly Canadian" with the CBC acting as the "distinctively Canadian" foundation of the system, has been so badly eroded through regulatory change and neglect that it is irrelevant and immaterial. [46] Where the CBC had once been seen as the vital cornerstone of the Canadian broadcasting industry, an exemplar of the responsibilities of all users of the public airwaves to foster Canadian culture and values, it has become an appendage, under constant threat of irrelevancy or total extinction.

Around the time the Broadcasting Act was being debated in Parliament and its committees, revolutionary digital compression technologies, which dramatically lowered the cost of cable and satellite distribution of TV signals, made possible an explosion of cable specialty channels. The broadcast regulator, the CRTC, awarded scores of licences for these new channels, almost exclusively to existing commercial broadcasters. Most were constructed on platforms already developed by American broadcasters and adapted, with some added Canadian content, to audiences here. HBO was repackaged as HBO Canada; Comedy Central became the Comedy Network; the History Channel became History Television; Bravo became Bravo Canada, and so on. They were expected to be highly profitable because development costs had been largely taken care of in the US, most of the programming was bought at an 80 percent discount, and most of them brought in revenue from both advertising and pass-through from cable subscriber fees.

Thus, while television audiences were badly fragmented by the many new viewing options, owners of both conventional television stations and cable channels were able to "reassemble" those audiences over multiple media platforms. Multimedia conglomerates like Bell Canada, Rogers, Shaw, Astral, and Quebecor, discovered that programming could be endlessly cross-promoted across newspaper, magazine, television, radio, and online platforms. However, these benefits were available only if corporate owners were big enough to assemble a large stable of such services, conventional and digital, capable of pulling together cumulative audiences large enough to interest advertisers; this prompted the scramble of buyouts and leveraged takeovers that has given Canada's media landscape its current claustrophobic, monopolistic structure. [47]

The CBC was largely sidelined from this transformative development, partly because of its perennial money woes, partly because of senior management who had little knowledge of the industry, but also because the CRTC refused those applications it did make — eight of them were turned down in the 1990s.[48] (Licences for cable news services in English and French were granted to the CBC in 1987 and 1995, respectively.) Thus not only was the public broadcaster denied access to the fountain of revenue represented by cable pass-through fees, but it had to cope with exponentially increasing competition for audiences in virtually all genres of programming in this new, highly-fragmented universe.

SUBSIDIES AND OPPORTUNITY

The picture of the current Canadian media landscape can only be sketched in outline here, but that is all that's necessary to make the case that, while the picture might seem to portray a hopelessly daunting challenge to the CBC, it is also a source of hope for the future. Political opportunity beckons.

First of all, given the current and ongoing state of ownership concentration in Canadian media,[49] the public broadcaster is quite clearly the only bulwark against monopoly control and its noxious effect on programming of all kinds. (I have excluded mention of newspapers for brevity's sake: the situation there is if anything worse than in radio and television.) This anti-monopoly, pro-market argument is one that can be made politically attractive within the current context of neo-liberal economic ideology, which sees monopolies as being fatal to the healthy functioning of markets of all kinds. Monopoly is anathema to laissez-faire liberalism.

Next, there is the issue of public subsidies to commercial interests. This is another area of interference in the free functioning of media markets that runs counter to liberal market theory. While a case can be made within liberal economic theory for public provision of services that the market cannot or will not provide (technically, "public goods"), subsidies to commercial interests operating in a competitive market amount to governments "picking winners and losers," another sin against neo-liberalism.

The Canadian broadcasting industry has been replete with such market-distorting government subsidies to commercial interests since the very earliest days of Canadian radio. In 1929 a Royal Commission of Inquiry established by the Liberal government of Prime Minister Mackenzie King and headed by John Aird recommended establishing a BBC-style monopoly for public service broadcasting in Canada, but history intervened, dramatically, in the form of the Wall Street crash and the election victory in 1930 of R.B. Bennett's Conservatives. A low-cost, compromise solution was adopted in which the forerunner of the CBC, the Canadian Radio Broadcasting Commission, would be set up to operate within the existing commercial milieu. The public broadcaster was permitted a small handful of publicly-owned stations, and coast-to-coast coverage was established through affiliate agreements with about a quarter of existing private broadcasters, on very profitable terms. Not only did the CRBC provide several hours of ad-free free programming a day, it paid for the wireline delivery charges, and it paid the stations to air it. The increased audiences allowed the stations to boost their rate cards for their own programs.

CRBC programs proved popular, and public pressure mounted for expansion of the service. But, as is consistent with public broadcasting's history in this country, the CRBC was badly under-resourced, and in 1935 it turned reluctantly to advertising to supplement its public funding. (The CRBC would be replaced by the CBC in 1936, as part of a general restructuring of broadcasting undertaken by the newly re-elected Mackenzie King Liberals.) Advertising fees did not reflect the true cost of the capital investment involved in establishing the coast-to-coast infrastructure, and so in making available a broadcast network for national and local advertising, the CRBC was providing another valuable service to private industry at public expense: a subsidy, in other words. As well, government investment in the CRBC and its successor contributed powerfully to the early technical and financial development of a thriving and highly profitable broadcasting industry, first in radio, and then in television. When television was launched in 1952-53, the CBC once again provided a backbone of programming for a rapidly growing number of private stations, making most of them instantly profitable. In those early years, private stations obtained an average of 60 percent of their programming from the CBC; in many cases the proportion ran to 85 percent. The subsidy continued well into the 1960s, at a cost to taxpayers of hundreds of millions of dollars (Weir, 262). More recently, the Conservative government of Brian Mulroney decided in 1984 to replace the Canadian Film Development Corporation with a new crown corporation called Telefilm Canada. Its job is to oversee public subsidies for Canadian content in both film and television programming, and in the process it has effectively changed the CBC from a program producer, to a commissioner of programs made by independent, for-profit production houses. Private producers welcomed the removal of their chief competition from the marketplace.

And the story continues. Between 1991 and 2009, while the federal government's expenditure on CBC/Radio-Canada rose by only eight percent, federal subsidies to private, commercial broadcasters rose by somewhere between 48 to 59 percent, depending on how the calculation is done. These subsidies came partly in the form of advertising-substitution regulations, which require broadcasters to replace American ads with Canadian advertising in US programming they purchase and simulcast on over-the-air transmissions. This CRTC policy is known as simultaneous substitution, or sim-sub. And when an American and a Canadian network are airing the same American programming at the same time, the policy requires Canadian cable and satellite television providers to substitute the Canadian broadcaster's signal for the American. This is why Canadians don't get to see the American Superbowl ads when they watch the game on Canadian television, and why ABC programming appears on CTV with the CTV logo embedded on the bottom right corner of the screen.

Sim-sub is a bonanza for the Canadian broadcasters, which procure programs for a tiny fraction of their production cost and charge full commercial rates within them. In 2010 the value of this subsidy was estimated to be between $182 and $204 million annually. [50] These numbers reflect the impact of the Great Recession; in 2007 the figure had been closer to $300 million,[51] and the trend is once again upward.

And then there's Section 19.1 of the federal Income Tax Act, which prohibits Canadian businesses from deducting, as an expense for income tax purposes, any advertisements they might place on foreign broadcasters that are "directed primarily to a market in Canada." The policy was implemented half a century ago to prevent American television and radio stations close to large Canadian urban centres from draining off advertising dollars with their cross-border broadcasts. The value of this federal subsidy to Canadian commercial broadcasters was estimated in 2010 to be between $92 and $131 million a year (Nordicity) and $120 million in 2007 (CRTC). Commercial broadcasters can also apply for program development and production subsidies from Telefilm, the federally-supported Canadian television production fund which pays out about $200 million a year. In a typical scenario, a network, say CTV, fronts the first 20 percent of production costs for a new series, and that triggers several sources of public monies: first, tax credits of 25 percent; and then funding of up to 15 percent from the Canadian Media Fund (run jointly by the Department of Canadian Heritage and the cable industry and administered by Telefilm); and up to 49 percent funding from Telefilm itself. Public subsidies thus can account for up to about 80 percent of the network's costs of production. The CBC is also eligible for this funding.

There's more: in 2007, the CRTC responded to pleas from private television stations hard hit by the Great Recession by setting up an emergency fund for the production of newscasts and other local programming. It was to be phased out in 2014. The money came from a mandatory tax on cable and satellite signal providers of 1 percent of revenue, a cost which most of these providers immediately passed on to their customers, so that the fund was in effect paid for by cable and satellite subscribers. In 2011, eighty local private stations received $106 million from this pool of money; over the life of the fund, it will have paid out roughly $600 million in subsidies.

The ostensible purpose of lavishing this government largesse on private television production is to encourage the commercial networks to produce more Canadian content to offset the hundreds of millions ($600 million in 2012) they spend on American programs each year. And it has done that. However commercial networks do what they do to make profit, and profit is maximized when this heavily-subsidized Can-con can also be sold in foreign markets. The result has been that many of the "Canadian" television programs produced with Telefilm money are scarcely recognizable as Canadian on the screen, a feature that makes them easier to sell abroad. The values they embody are not so much Canadian as commercial. Furthermore, commercial networks tend to air their Canadian content in off-peak viewing hours, leaving prime time to their more lucrative American imports.

All in, federal subsidies to Canadian commercial broadcasters total something in the neighbourhood of $300 million dollars a year for the sim-sub and tax-based subsidies combined. With various other tax incentives for Can-con film and video production, the local programming fund, and Telefilm subsidies, the total, though difficult to calculate accurately, would appear to be within shouting distance of the federal subsidy to the CBC, which in 2011 was about $1 billion.

The question this raises is: why? Why should Canadian taxpayers subsidize a handful of highly profitable media conglomerates whose television networks make their money by rebroadcasting American television shows purchased at about five percent of their cost of production, and which relegate their mandatory Canadian content quotas to newscasts and off-peak viewing hours?

A NECESSARY BALANCE

At a time when governments, both federal and provincial, are slowly emerging from the financial crisis of 2008, and when the public broadcaster is in peril of being brought to its knees by its ill-advised involvement with professional sport and its dependence on advertising, it has become undeniably clear that something must be done to preserve balance in the nation's media ecology.

We live in an era of epic challenges, both national and global, arising out of financial instability, geopolitical transformations, environmental calamity, massive migrations of desperate human populations, religious animus, terrorism, the threat of nuclear weapons, runaway developments in science and technology. That much has become conventional wisdom. What is not so well understood is that, for these reasons and many others, there has never been greater need for the kind of thoughtful dialogue and considered judgment that can take place only in public spaces. The most important of these is provided by our media.

There was a time when broadcasting was understood by both the public and by its entrepreneurs to involve a degree of public service, thought of either as a moral responsibility or simply a quid pro quo for permission to use the public airwaves. That era of media came to an end beginning in the 1980s with the rise of neo-liberalism, and the messianic belief in the moral authority of unfettered market capitalism. If we can judge from such manifestations of social disaffection as the Occupy movement, we now appear to be entering a phase of reaction to this, of regret over the destruction of so much social value embodied in the public service initiatives of the great commercial media organizations of the middle decades of the twentieth century. It seems highly doubtful, however, that the clock can be turned back where the mammoth corporate media conglomerates are concerned — they have become too big and powerful and politically potent to be persuaded to forego profit in favour of a renewal of their public service mission. No professionally managed television network, for example, will voluntarily decide to restore its spending on news and current affairs to pre-deregulation levels simply because it is socially responsible; "the right thing to do." The laser-like focus in these organizations is on their fiduciary responsibility to shareholders (i.e., generating profit).

Clearly, the public service *assumptions* (though not the prescriptions) of Canada's Broadcasting Act of 1991 are outdated. But if we can no longer rely on commercial media to serve the public interest beyond some minimal entertainment function, we are fortunate to still have more than a remnant of true public service broadcasting in the CBC/Radio-Canada. We have a strong rootstock on which to begin to grow a new system more suited to the media ecology of the twenty-first century as defined by our current political-economic, social, and technological realities. If we can no longer depend on commercial broadcasters to educate and enlighten in any significant way, we can at least insist that they entertain us responsibly, and in this country there remains a body of regulation administered mainly by the CRTC that prevents broadcasting from descending to quite the depths plumbed in some of the news commentary and reality programming seen in the US. There are no equivalents to *Extreme Makeover* or the *Rush Limbaugh Show* being perpetrated here: to do so would put broadcasting licences in jeopardy. The CBC is entirely capable of filling the public interest gaps in the system. What it needs is a sustainable, predictable, level of public funding and a management team that understands and supports the goals of public service media and is willing to do what it takes to achieve them. All of this can be realized, if we can first achieve public recognition that the provision of public space for the crucial debates that will shape our future is too important to be left in the hands of a small coterie of profit-driven corporate oligopolists.

With that in mind, I offer the following proposals in the interest of furthering the urgently-needed debate on the future of public broadcasting in Canada.

9

Ten proposals for a new CBC

The suggestions that follow have, for the most part, the support of a broad swath of opinion among people who care about the future of the public broadcaster, from members of watchdog groups like Friends of Canadian Broadcasting to former and current employees of the corporation. What I have tried to do in the preceding chapters is provide a strong body of fact and theory on which an action plan can be based. These proposals constitute that plan, if only in broad outline. It is a plan I believe to be achievable, and which can ensure the long-term survival of the CBC.

Proposal 1

Eliminate advertising on CBC television and prohibit its reintroduction on CBC radio. This is fundamental to the revitalization of the CBC as a public service broadcaster. The approximately $400 million in ad revenue collected annually by the CBC would then be available to private broadcasters who have for nearly a century complained bitterly about competition for advertising dollars from the publically-subsidized CBC and its predecessor, the CRBC. Give them what they want, at long last.[52]

One of the reasons private broadcasters resent the CBC is that the corporation can undercut them in competition for advertisers, thanks to its subsidy. If the CBC were to end commercial sponsorship, private stations would see a substantial increase in revenue both from increased ad rates and reductions in unsold inventory. Furthermore, according to the CRTC, the CBC spends $136 million a year on sales and promotion: industry insiders estimate about $100 million of that goes to support its advertising sales department with its hundreds of agents.[53] This is money that could be spent on programming.

Proposal 2

In return for providing substantial new advertising revenue to private broadcasters, increase the CBC's $1.1 billion parliamentary appropriation by an equivalent amount, to be taken from the various subsidies currently provided to private broadcasters. The object is to provide stable, multi-year financing for the public broadcaster, at a level that will permit it to properly fulfill its mandate for public service on both television and radio, and on the web. The details of financing will obviously be complex and will require new legislation. The guiding principle should be that the public service broadcaster be treated as a utility — the provider of a public good that is universally accessible regardless of income or geographic location; a necessary service paid for from the public purse, and not through charitable donations, or subscription fees, or bake sales.[54]

Proposal 3

Eliminate professional sports broadcasting on CBC. Proposal 1 demands this. Pro sport, typified by the NHL (a $3 billion industry), is big business, and as such demands very high prices for the television content it provides. No broadcaster could afford any of the big pro sport franchises without offsetting advertising revenue, so, carrying sports means, for the CBC, advertising on television. The CBC's withdrawal from sports will provide a revenue windfall for private broadcasters, a portion of which should be transferred back to the CBC to supplement the Parliamentary appropriation.

Beyond financial considerations, the positive values sport embodies are to be found most authentically in amateur sport, and not here. Whatever the NHL may have been in its infancy and adolescence, the mature, American-dominated version can no longer be said to represent Canadian values. It has degenerated, for the most part, into often violent spectacle and a culture that places money above almost any other consideration, including the welfare of the team and hometown pride. The values it represents are no longer social, but commercial. *Hockey Night in Canada* began its life in 1931 as an idea cooked up for clients by the MacLaren Advertising Agency; for many years it was known as the Imperial Esso Hockey Broadcast. It has always been, in its essence, a promotional vehicle for its sponsors and for the NHL itself. It nevertheless remains popular as entertainment, and no one would suggest that Canadians be denied the pleasures of watching hockey, or even of enjoying the tradition of the *Hockey Night in Canada* format. Its natural home today is on commercial networks, which specialize in providing amusement and relaxation without much thought to values. Over the years CBC has lost the rights to the Canadian Football League, Major League Baseball, the National Football League, and other pro sports to specialty channels like TSN, Sportsnet, and the French-language RDS. The sports specialty channels collectively broadcast three times as many hockey games as CBC during the season, although many of their games involve contests between American teams. The CBC broadcasts about 100 games, and has exclusive rights to the playoffs. (CBC/SRC in Quebec relinquished rights to NHL hockey in 2001 to the sports specialty channel RDS, with no apparent negative impact on the public network's popularity.)

The CBC keeps the cost of its NHL franchise secret, but analysts estimate it to be around $100 million a year in the current contract. Add to that another approximately $25 million for production costs, including salaries for on-air talent Don Cherry and Ron MacLean, for a rough total of $125 million.

According to Barry Kiefl, head of research at CBC until 2001 and now an independent media research consultant (on whose work most of the figures in these paragraphs are based), NHL hockey brings in somewhere between a half and a third of total CBC television advertising revenue, which fluctuates from year to year but in 2011 was about $250 million. So hockey brings in about $100 million in advertising revenue each year; readers will have noticed that's less than the production and contract costs of $125 million.

But that's not all, Kiefl reports:

> The true cost of HNIC can't be measured without considering the cost of selling the ads in hockey and other CBC programs. CBC's sales and promotion expenses have increased markedly since 2008, from under $50 million to over $80 million. Some of the $80 million that CBC spends on sales ... is spent on HNIC, perhaps $20-$30 million.... Adding in the cost of sales means CBC's HNIC is losing tens of millions of dollars with the current NHL contract, perhaps as much as $50 million. (Sept 15, 2012) The CBC's continuing involvement with the NHL makes no sense, either financially or in terms of public service. Canadians will not be denied their hockey if it's not on the CBC. And it is clear to anyone who pays attention to the CBC on radio and television how the irrational dependence on NHL revenue distorts the corporation's public service values by, for example, preempting evening newscasts for playoff games, and cross-promoting NHL hockey at every opportunity as the "national sport."

The 350-hour "hole" in the network's Canadian content schedule that would be created by the loss of hockey can be filled at an affordable cost with creative programming initiatives. As Kiefl says: "...the CBC would be wise to think about programming strategies that do not include NHL hockey, as Radio Canada did years ago, which made SRC a distinctive and more vital service. Initially there was a public (read: media) outcry and the CBC president was hauled before a Parliamentary committee to explain this horrible decision but within weeks it was all but forgotten."To ensure that the commercial broadcasters who will inherit nationally important sporting events do not restrict access by placing them on money-spinning subscription-only cable channels, the CRTC should, after consultation, make a list of events that must remain public, i.e. available on over-the-air channels or basic cable. Obvious candidates for this list are the Stanley Cup playoffs, the CFL's Grey Cup playoffs, and the Olympic Games.

Proposal 4

Put the CBC back in the business of producing television programming, rather than merely commissioning and purchasing it. This would stimulate the market for authentic Canadian programming in which Canadians can see themselves as who they are, rather than as stand-ins for some other nationality in some other location.

The slimming down of the CBC from broadcaster-producer to broadcaster-publisher in 1984 was justified as a way to introduce greater diversity of voices in program offerings, to support and help develop the independent film production industry, and above all, to save money. Save money it did, by forcing artists and craftspeople of all descriptions into the precarious world of independent film-making, where employment is sporadic, wages are relatively low, and job security and benefits are a chimera. Ignoring the moral issues raised, it is possible to ask whether or not there is a threshold in size and what might be termed "creative density" beyond which an enterprise like the CBC cannot pass without an irretrievable loss of identity and functionality.

Commenting on efforts in the UK to reduce or eliminate in-house productions at the BBC, media scholar Michael Tracey has written: "The key question here is whether or not in slimming down the large public broadcasting organization, for example, by introducing internal markets and producer choice and allowing funds to flow outside the organization, one triggers a kind of cultural anorexia. There is an important, but highly abstract and intangible, argument that successful public broadcasters tend to be 'largish', with sufficient creative mass to fund, nurture, and give space to talent across a range of genres. Shrink that size too far and the institution becomes impoverished." (266)

Proposal 5

The senior management of the CBC has been rendered dysfunctional by political patronage and needs to be restructured in a way that will put it in the hands of committed public servants who understand and are sympathetic to the role of public service broadcasting in the nation's media ecology. The guiding principle for management should be democratic representation in the realm of culture.

The CBC president is currently appointed by the federal Cabinet. For the past fifty years, these appointments have been drawn almost exclusively from the senior ranks of federal civil service, with the overwhelming emphasis being on men (they have all been men) who had a firm grasp of accounting. As the veteran producer Richard Nielsen has written: "Canadian broadcasting, like all broadcasting, is an extremely complex industry involving technological innovation; business arrangements that involve foreign countries; production involving sometimes difficult and always challenging artists and producers, and distribution patterns that are constantly changing, all of it involving a rapid pace of technological change. Canada alone seems to have decided that no experience whatsoever is required in conducting and leading such a corporation." He adds, ruefully: "I've known and had dealings with practically all the Presidents beginning with Alphonse Ouimet [appointed 1957] and they were honourable and intelligent men who with few exceptions left their posts as ignorant as they had arrived, proving that Ottawa is not a good vantage point from which to understand the industry nor is it an industry that can be learned from the top." (8) Successive federal governments have seen the role of President of the CBC as involving principally curbing a perceived tendency to lavish spending, and, from time to time, providing a counterweight in Quebec to separatist sentiments.

If the CBC is to be allowed to do its job at arm's-length from government and its political preoccupations, the president should be chosen by the corporation's Board of Directors, and should serve at its pleasure.

As to the Board of Directors itself, for the past half century and more, federal governments have made appointments more on the basis of political alignment than knowledge of the industry or commitment to the values of public broadcasting.[55] There needs, therefore, to be a system for appointing board members that is at arm's-length from government. In the UK this is accomplished by a process designed by the distinguished jurist Lord Nolan in 1995 at the request of the Conservative government of John Major, in response to a patronage and corruption scandal. The Nolan Rules, as the system has become known worldwide, is a non-partisan selection process that provides for transparency, careful assessment of merit, and oversight by a Commissioner of Public Appointments (who is independent of both civil service and government), in the process of selecting a short list of candidates from which the Minister then makes his or her choice. The system is designed explicitly to eliminate the kind of patronage appointment that has characterized the CBC's board for generations. Some version of it could and should be, adopted here.

A Board of Directors (currently ten persons, plus the President and the Chairman, all appointed by the Cabinet) selected under Nolan Rules would elect a Chair from its own ranks, and would then hire a full-time President, or chief executive officer for the corporation.

Proposal 6

Find ways to improve public accountability, so that, for example, decisions to alter programming or services do not come as a surprise to the audience, making them feel disenfranchised and alienated. This was a lesson that should have been learned with the disastrous 1990s decision to abandon local and regional programming and services in many areas of the country. Public consultations could have averted or mitigated the impact of this blunder. CBC is not a private company, and it is only its current competitive position in the advertising market that justifies the secrecy surrounding its decision-making (such as how much it pays for NHL rights). Once out of advertising, it will be in a position to involve its audience in making the decisions that affect the services they rely upon and care about. This can only improve the quality of those decisions, while at the same time cultivating loyalty. There are many ways to accomplish this: one is to establish a foundation for the support of public broadcasting that would be charged with this responsibility. The foundation could also be a high-profile champion for the values of public service media.

Proposal 7

Expand CBC radio service to include at least one more over-the-air network, which will focus on classical music, the arts, and drama. It was a mistake to alienate the most loyal of CBC listeners, those who were devoted to Radio 2 prior to its transformation into an "adult contemporary" format in 2009. At the same time, there is undeniable logic in the provision of an outlet for Canadian popular music that seldom is heard on commercial radio. The country needs, and deserves, both, in addition to the currently very popular and necessary Radio One. Yes, there are commercial radio stations that play classical music in some markets, but that's all they do, and they don't present it as intelligently as CBC Radio 2 did. That is because they do it for a different reason: not to educate and enlighten, but to sell products for advertisers.

The CBC's array of streaming online music channels (www.cbc.ca/music) is impressive and useful as a supplemental service, but it is no replacement for broadcast services that bring large audiences together and offer intelligent curation and presentation of musical offerings. One is a record store; the other is a combination concert hall and musicology course.

Proposal 8

Expand CBC television on cable (and perhaps on the streaming web-based subscription services like Netflix that appear to be replacing traditional cable carriage). CBC must maintain its current over-the-air television services, but focus future expansion on digital "specialty" channels. These services should be part of all cable and satellite providers' basic packages, i.e. mandatory carriage. Premium subscription services (or PPV — pay-per-view) are incompatible with the CBC's mandate to provide a balanced service to all Canadians. The Pilkington Committee report on public broadcasting in the UK addressed this point definitively half a century ago: "To finance the Corporation in whole or in part from the proceeds of the sale — to those who want and can afford them — of particular program items would...positively discourage and make more difficult the provision of a balanced service. We reject, therefore, as opposed to the purpose of public service broadcasting, the idea that the BBC should engage in subscription television."[56] Even basic cable is an unaffordable expense for many Canadians, and as over-the-air signals are eliminated in the move to high-definition television, ways must be found to ensure that the public broadcaster remains universally available, perhaps through a tax-based rebate system, or a system of vouchers. The same argument applies to broadband delivery via the Internet.

The CBC's current "bold" channel, despite its peculiar name, lower-case affectation and lack of CBC branding, carries some interesting re-runs from the UK and the US ("Mad Dogs," "Skins"), reruns of Canadian series, as well as video recordings of popular CBC radio shows like "Q" and "Saturday Night Blues," and provides the basis for further development as an outlet for original alternative programming. But it needs to be on basic cable and advertising-free in order to explore its full potential.[57] The same goes for "documentary," with its all-documentary-all-the-time format, a joint venture of the CBC and National Film Board of Canada, although for some reason it is not branded this way. These channels each generate only about $100,000 in advertising revenue, which is about seven times less than their respective sales and promotion budgets.[58] Repositioning them as ad-free would save money.

CBC News Network on cable provides an essential, and valued, service, but it can reverse a slow but steady down-market slide and fully explore its potential only if it is freed from the strictures of advertising, which, among other things, forces it to shape news coverage to fit advertising schedules. Time matters: although North American audiences have become accustomed to commercial interruptions in news coverage, the fact remains that advertising cheapens and trivializes news, drawing even the most important and significant stories into the realm of consumer culture through juxtaposition. News is not just a product to be consumed and should not be treated that way, not, at any rate, on the public broadcaster.

Advertising revenue for News Network and its French-language counterpart, RDI, is more substantial than for the other cable channels, but still amounts to less than five percent of overall CBC television advertising revenue. The value to viewers of an advertising-free, truly distinctive all-news channel would seem to far outweigh this. A tiny increase in cable pass-through rates to the CBC would cover the loss.

Proposal 9

Continue to expand web-based, digital services, but do it thoughtfully. The CBC and the rest of us need to remember that a broadcast is something more than a podcast or a streaming audio or video feed — it is the assembly of a congregation, to use Richard Nielsen's words. Online, computer-accessed services are no replacement for over-the-air presence. You can't listen to or watch your computer or mobile phone while driving your car. You can't even listen to online streaming radio or podcasts on your home entertainment set-up without dealing with a technical interface as complicated as the first tube-type radio receivers with their plug-in coils, batteries, and forests of dials and knobs. The radio industry figured this out in the 1930s and gave us the simple, two or three knob plug-in box that fit on a window ledge and could be operated by a small child. As well, the idea that listeners who manage to wire up their computers to their stereos should then have to pay bandwidth charges to their Internet service providers (ISPs) for every minute of radio or television they consume, is in direct opposition to the basic, universal accessibility premise of true public service broadcasting. Not everyone has a computer; not everyone can figure out how to make it work through a decent sound system; not everyone can afford bandwidth or data usage fees charged by ISPs.

Proposal 10

Restore or initiate local television and radio service in as many medium to small-sized communities as possible. Where necessary, do this in collaboration with independent community-based media outlets. The

CBC's support for these small but dedicated non-profit organizations will be a boon for them, and they will provide CBC with a widening recruiting pool of tested talent for the radio and television networks.

Commercial television networks have found that a local early evening newscast is a key component to assembling a large and profitable audience for an evening of viewing of the advertising-rich American dramas and sitcoms that follow. And local newscasts, with their well-groomed and well-promoted presenters are an important component of a station's identity, helping it to develop viewer loyalty in much the same way as the national newscasts fronted by Cronkite, Jennings, Chancellor, Brokaw, Kirck, and Robertson were at one time thought of as the public relations standard-bearers of their respective networks.

CBC learned the hard way. In the 1990s CBC managers chose to respond to (yet further) draconian budget cuts by virtually eliminating the network's regional television services, pulling back from local television news coverage, and closing down eleven stations altogether. This proved to be a disastrous strategic error. It provoked an angry public outcry of startling ferocity, and disaffected audiences in these markets deserted network programming across the entire broadcast schedule. The CBC's thinking at the time had apparently been that if budgets had to be cut, Canadians would be best served if cutbacks were made in areas being well-served by commercial media, and from this perspective local television news seemed an obvious choice. The commercial network affiliates were there, and doing a reasonably good job.

But it is a mistake to assume that commercial television outlets will maintain high standards where competition from the public broadcaster is removed. Their interest is in preserving audience numbers, and how this is done is of secondary importance. Over time, this means that news programs devolve into infotainment, fronted by attractive, focus group-tested presenters and reporters, and emphasizing crime, weather and celebrity — all of which is cheaper to produce than real journalism, but still generates acceptable ratings. Or, the programs may simply be scuttled.

In recent years CBC news, for all its problems, has been trying to rectify the damage done by its withdrawal from local service in the 1990s. But, of course, it no longer has the funding necessary to do the job properly. Meanwhile, ongoing mergers and acquisitions in commercial television had by 2010 raised debt levels to a point where corporate managers, facing the economic slump, saw a need to downsize or eliminate long-established local news operations in smaller stations across the country. Their decision to do so caused considerable pushback from the audiences to be affected, to little effect.

Conclusion

Advertising was introduced to broadcasting in this country with great reluctance, as it was in the United States. In both countries, the 1920s was a time of progressive politics, and the idea of public service was an active meme in both business and politics. In America, the resistance to commercials on the airwaves was quickly overcome when the technology was created to string together nation-wide networks of radio stations over long-distance telephone lines, dramatically lowering programming costs and opening up the country to national advertisers. The revenue potential was simply too great to be ignored. But the network entrepreneurs were creatures of their time — men concerned with their good name and willing to trade profit for prestige. Network advertising was confined almost exclusively to sustaining sponsors, who paid for entire programs and series in order to have their brand attached to popular, high-quality entertainment. Out of this arrangement came the excellence nowadays characterized as the golden age of radio. It ended with the advent of television.

In Canada, commercial sponsorship was adopted reluctantly as the only expeditious way to build a national broadcasting network and provide the kind of high quality programming enjoyed by listeners south of the border. Successive federal governments and CBC administrations have characterized the "blended" commercial/public service system in this country as a brilliant compromise, a win-win. Throughout most of the history of Canadian broadcasting, the CBC functioned both as a stand-alone public service broadcaster, and also a producer of public service (read: Canadian) programming which it supplied to private stations at no cost, as compensation for their willingness to have taken on the risk of setting up radio and television stations throughout the country. The "risk" for many was minimal: the business of regional broadcasting was famously characterized as "a licence to print money," as private broadcasters loaded up their schedules with cheap American programs and filled blank spots with free CBC fare. Nevertheless, government support for private, commercial broadcasting in Canada has continued to be a significant part of the institutional structure of the broadcasting industry. This no longer makes sense, given the realities of the industry today.

Private broadcasters are no longer local, independent businessmen struggling to make ends meet while providing needed services like local news and weather, community information and entertainment to their communities. Private broadcasters of today are typically enormous corporate conglomerates like Rogers and Bell and Corus and Astral and Power Corp. and the handful of other corporations that own and control Canadian commercial broadcast media. Commercial broadcasting has become an industry like any other — dominated by widely-held, professionally managed corporations that are narrowly focused on maximizing shareholder value. It can no longer reasonably be considered to be preeminently a public service; it serves advertisers and shareholders. The historical record shows that, with few exceptions, pioneer private broadcasters could be counted on to display a sense of public responsibility, because their personal reputations were at stake. But nobody expects modern corporations to behave this way: their sense of responsibility to the public, as opposed to their shareholders, begins and ends with compliance with law and regulation. If we want private broadcasters to serve public policy ends such as including a certain level of Canadian content in their broadcast schedules, the responsibility must be imposed through regulation. But even then, it is naïve to expect any large, publicly-held, and professionally managed corporation to do anything more than comply with the exact letter of the law, rather than its spirit. This makes the drafting of regulations, and their enforcement, exceedingly difficult and often politically unpalatable.

Only a public broadcaster will serve the public interest wholeheartedly and with imagination, creativity, and real enthusiasm. That's why this country — so intimately linked through communication technologies of all kinds to the world's biggest exporter of cultural products — needs the CBC if it is to be all that it can be. We are in danger of losing it, at least in any recognizable form.

Complex organizations of any kind need to maintain a certain size if they are to survive and flourish. Underfunded players in competitive markets tend to be robbed of their most valuable employees, either because they can't offer competitive salaries, or because opportunities are limited in comparison with rival firms. There will come a point, if it has not already been reached, when the CBC will be so starved, so emaciated, that its best, most creative employees will give up the ghost. Ivan Fecan, recently retired CEO of CTVglobemedia, is a prime example of the process, lured away in 1994 from his position of Vice-President, English language television, at CBC to CTV by a lavish compensation package and, rumour has it, a shiny new Mercedes convertible as a signing bonus. He quickly became the CBC's nemesis in regulatory proceedings, and negotiations for hockey and Olympic franchises. Norm Bolen left his position of head of CBC television current affairs programming to run History Television for Alliance Atlantis in 1997, eventually becoming CEO of the Canadian Media Production Association. Tony Burman left his position as head of television news for a senior position at Al Jazeera. Jeffrey Dvorkin left his post as head of CBC radio news for a position with NPR in the US. Slawko Klymkiw left as head of English language television programming to become CEO of the Canadian Film Centre – the list could go on for pages. There remain many hundreds of bright, talented, and dedicated people holding the fort at CBC production centres across the country and in a few remaining bureaus around the world. But morale is not good, and their prospects are, to say the least, uncertain.

There will come a point when further cutbacks to the CBC's funding will no longer lead to quantitative tinkering with its output, but to fundamental, qualitative transformation in the organization itself. I am among a large number of knowledgeable observers who believe that stage will be reached within the next two years. The tipping point will in all likelihood be the loss of NHL hockey and its associated revenue, but it could be an number of other potential financial disasters, including further cuts to its Parliamentary appropriation. There is an urgent need for public debate on saving public service broadcasting in Canada and restoring it to health.

Think of a world in which all roads are toll roads; in which city parks charge admission; in which all medical care is provided for profit; in which there are no public schools and universities, just private academies; in which there are no public museums of art and history, no public libraries, just private galleries and bookstores; in which shopping malls provide our only "public spaces." Now think of a world in which all media of information and entertainment are for-profit ventures that provide content based solely on its ability to attract large audiences of consumers.

Do we want the latter world any more than we want any of the former?

The CBC has been, and can be once again, the crowning achievement of the nation's desire to provide a rich and fulfilling cultural life for all of its citizens; to provide the means by which Canadians of every background and condition can connect with one another for purposes other than buying and selling — in order to know one another better, and to further their common interests through the alchemy of informed, civil communion in public space.

Notes

1. From 1985 to 2010, the CBC's Parliamentary appropriation — its taxpayer subsidy — went from $905 million to $1.018 million, a nominal increase of 12.5 percent, but an actual decline of 62 percent after inflation. In 2011 the federal government imposed a further 15% reduction to the CBC subsidy over three years.
2. Among the 26 OECD nations, Canada's subsidy to its public broadcaster, as a percentage of GDP, ranks 22nd; it is about half the average, and less than one-third that of countries such as the United Kingdom, Norway, and Denmark. At the same time, polling has repeatedly shown that Canadians value the CBC highly, and would be willing to pay substantially more for service if given the opportunity. In 1998 Parliament's Standing Committee on Canadian heritage recommended that the CBC's annual public subsidy be set at a minimum of $40 per capita, about eleven cents a day; it is currently less than half that.
3. I make the assumption throughout that television is, and will remain for the foreseeable future, the dominant medium of information and entertainment, worldwide. The flourishing of digital, web-based media notwithstanding, it is television (whether viewed on conventional sets or on mobile digital appliances) that mainly shapes our view of the world and its denizens, and meets most of our entertainment needs. The most recent comprehensive survey of media use in an advanced industrial nation was undertaken by Britain's Ofcom regulator in 2010. It found that people spend more time engaged

with various media (radio, TV, computer-based devices) than in any other activity including sleep: about 40 percent of the waking day was spent with media as a sole activity or in addition to some other activity. Conventional television viewing continues to dominate media use in evenings. Americans currently watch about 34 hours of TV a week; Canadians watch about 27 hours a week (Nielsen). Radio, despite dire forecasts following the rapid uptake of TV in the 1950s, retains a prominent place in our lives, now mainly as background soundtrack (except on public broadcasters), precisely because it has been able to exploit its niche as the entertainment and information medium that lets you get on with our life. It's a valued companion.

4 Remaining provincial public broadcasters are TV Ontario and its French-language equivalent TFO, British Columbia's Knowledge, and Télé-Québec, which is government-owned but advertising-supported. Saskatchewan's SCN network, a strong regional presence on television, was privatized in 2010 by the government of Premier Brad Wall and eventually sold to Rogers Media. "SCN's viewership is quite low," Dustin Duncan, the Minister of Tourism, Parks, Culture and Sport, said at the time. "We feel that there is no longer a role for government in the broadcast business."

5 The UNESCO publication "Public Broadcasting: Why? How?" defines public broadcasting as follows:

"Neither commercial nor State-controlled, public broadcasting's only raison d'être is public service. It is the public's broadcasting organization; it speaks to everyone as a citizen. Public broadcasters encourage access to and participation in public life. They develop knowledge, broaden horizons, and enable people to better understand themselves by better understanding the world and others. Public broadcasting is defined as a meeting place where all citizens are welcome and considered equals. It is an information and education tool, accessible to all and meant for all, whatever their social or economic status. Its mandate is not restricted to information and cultural development-public broadcasting must also appeal to the imagination, and entertain. But it does so with a concern for quality that distinguishes it from commercial broadcasting. Because it is not subject to the dictates of profitability, public broadcasting must be daring and innovative, and take risks. And when it succeeds in developing outstanding genres or ideas, it can impose its high standards and set the tone for other broadcasters. For some, such as British author Anthony Smith, writing about the British Broadcasting Corporation – seen by many as the cradle of public broadcasting – it is so important that it has "probably been the greatest of the instruments of social democracy of the century." (7) [http://www.unesco.org/new/en/communication-and-information/resources/publications-and-communication-materials/publications/full-list/public-broadcasting-why-how/]

6 The 1932 Act created the Canadian Radio Broadcasting Commission, which not only broadcast on its own, but regulated private broadcasting. The CRBC was abolished and replaced by the Canadian Broadcasting Corporation (a Crown corporation), and an independent regulator, in the Broadcasting Act of 1936.

7 The lack of coverage of the rest of Canada on SRC is striking: see the intervention of Senator Pierre De Bané, "The Trouble With Radio-Canada," CTRC Nov. 17, 2012, CBC/SRC licence renewal hearings. [https://services.crtc.gc.ca/pub/ListeInterventionList/Documents.aspx?ID=174810&Lang=e]

8 The Canadian Association of Broadcasters (CAB), the private broadcasting industry's chief lobbyist, was founded in 1926. A record of its opposition to the CBC can be found in E. Austin Weir's book *The Struggle for National Broadcasting in Canada*. (1965).

9 Gresham's Law was formulated in the sixteenth century by Sir John Gresham, who noticed that when coins containing different amounts of precious metal have the same face value as currency, shopkeepers and others will tend to hoard the more valuable coinage while using the less valuable for payment and making change. Thus, the "bad" coinage drives the "good" out of circulation.

10 One need not look to other countries for evidence of what is possible on television: the CBC's own schedule for 1960-1 provides an insight into what used to be considered normal fare on the network. *Festival* in that year presented Shakespeare's *Julius Caesar*, an adapta-

tion of Dickens' *Great Expectations*; a Stratford Festival performance of *H.M.S. Pinafore*; the operas *Electra* and *Falstaff*; Eugene O'Neill's *The Great God Brown*; Jean Anouilh's *Ring Around the Moon* and *Colombe*; S. Ansky's *The Dibbuk*; Oscar Wilde's *Lord Arthur Saville's Crime*; James Reaney's *The Kildeer*; Emlyn Williams' *Night Must Fall*; Henry James' *The Pupil*. The new program *Q for Quest* hosted by Andrew Allen presented in the same season interviews with Mordecai Richler, Maureen Forrester, James Reaney and others, and performances of Len Peterson's *Burlap Bags*; Brecht's *The Great Scholar Wu*; Chekhov's *For the Information of Husbands*; Theodore Bikel and Mary Martin in *The Sound of Music*; *The World of S.J. Perelman*; a performance by Lambert, Hendricks, and Ross; Jean Cocteau's *The Human Voice* and much more. In the season of 1956-7 "L'Heure du Concert" presented fourteen operas including Stravinsky's *Oedipus Rex* and Gounod's *Mirielle*, plus six ballets, including the National Ballet's *Swan Lake*. Also in that year, the English television network broadcast 48 dramas a week, more than half of them written by Canadians. *Folio* scheduled twelve plays, half by Canadian writers; *General Motors Theatre* presented eighteen hour-long plays including Arthur Hailey's *Flight Into Danger*; *On Camera* presented another 29 half-hour plays, most of them Canadian. [Weir 393ff.]

11 The CBC under current management apparently does not consider product placement to be advertising. CBC Policy 1.3.8 states: "The CBC/Radio-Canada does not accept advertising of any kind in programming and websites designated by the CBC/Radio-Canada as directed to children under 12 years of age. Products that appeal to children and in their normal use require adult supervision may not be advertised in station breaks adjacent to children's programs."

12 The preamble of the UK "Communications Act 2003" states that: "The public service remit for Channel 4 is the provision of a broad range of high quality and diverse programming which, in particular:

- demonstrates innovation, experiment and creativity in the form and content of programs;
- appeals to the tastes and interests of a culturally diverse society;
- makes a significant contribution to meeting the need for the licensed

public service channels to include programs of an educational nature and other programs of educative value; and
- exhibits a distinctive character."

13 He was hired by Robert Rabinovitch in 2004. Rabinovitch in a recent interview said of his tenure as CBC President: "I approached the job as a businessman. I looked at how our assets could generate money. I sold Newsworld International, I leveraged our real estate holdings. The bottom line is that I did not suffer from losses. I created more cash flow, which allowed us to enhance the quality of our programs, take more risks, try more series, update the delivery system to a digital model — and in the long run, save more money." Interview with Greg O'Brien.

14 The claim is exposed as Orwellian even within its own narrow terms of reference in a 2012 Canadian Media Trends research report, "The CBC, ex-CBC Executives and 'Factortion' [http://mediatrends-research.blogspot.ca/2012/04/cbc-ex-cbc-executives-and-factortion.html]

15 Technically, it's a logical fallacy called a tautology: the conclusion is embedded in the premise. Consider this as an exact parallel: Why do kids eat candy? Because it's good. How do we know it's good? Because kids eat it. Note as well that, in equating quality with metered popularity, those who share this view are stealing a page from early twentieth century analytic philosophy, which dealt with sticky problems of morality and aesthetics by simply declaring that they had no objective existence — that all such judgments were purely subjective and emotional. There was, in other words, no issue to be dealt with. The view has long since been discredited and consigned to footnotes in contemporary philosophical inquiry.

16 Todd Gitlin, 27. He adds, "Network executives often say their problem is simple. Their tradition, in a sense, is the search for steady profits. They want above all to put on the air shows best calculated to accumulate maximum reliable audiences. Maximum audiences attract maximum dollars for advertisers, and advertiser dollars are, after all, the network's objective. (Network executives recite the point so predictably, so confidently, they sound like vulgar Marxists)" (21).

17 The top 30 (of 50, compiled January 2010): *The Sopranos* (HBO); *Brideshead Revisited* (ITV, 1981); *Our Friends in the North* (BBC, 1996); *Mad Men* (AMC, 2007-); *A Very Peculiar Practice* (BBC, 1986-8); *Talking Heads* (BBC, 1988, 1998); *The Singing Detective* (BBC, 1986); *Oranges Are Not The Only Fruit* (BBC, 1990); *State of Play* (BBC, 2003); *Boys from the Blackstuff* (BBC, 1982); *The West Wing* (NBC, 1999-2006)); *Twin Peaks* (ABC, 1990-1); *Queer as Folk* (Channel 4, 1999-2000); *The Wire* (HBO, 2002-8); *Six Feet Under* (HBO, 2001-5); *How Do You Want Me* (BBC, 1998-9); *Smiley's People* (BBC, 1982); *House of Cards* (BBC, 1990); *Prime Suspect* (ITV, 1991-2006); *Bodies* (BBC, 2004-6); *Tinker, Tailor, Soldier, Spy* (BBC, 1979); *Buffy the Vampire Slayer* (The WB/UPN, 1997-2003); *Cracker* (ITV, 1993-6); *Pennies from Heaven* (BBC, 1978); *Battlestar Galactica* (Sci-Fi/Sky 2003-9); *Coronation Street* (ITV, 1960-); *The Jewel in the Crown* (ITV, 1984); *The Monocled Mutineer* (BBC, 1986); *Clocking Off* (BBC, 2000-3); *Inspector Morse* (ITV, 1987-2000). In the international reader comments that followed the posting, the program most referenced as should-have-been-there was another BBC production, *I, Claudius* (BBC, 1976).

18 Britain's commercial ITV system, which has produced programming of exceptionally high quality over the years, can be considered a special case due to the onerous public service stipulations attached by Parliament to its operating charter, and to the unusually high benchmark established by its competition, the BBC. The statutory public service requirements had the effect of ameliorating the impact of Gresham's Law; the BBC's well-established standards had produced a discriminating audience pool resistant to cheap, shoddy programs.

19 See Edward Herman and Noam Chomsky for the classic exposition of news "filters."

20 Leslie-Anne Keown, 2004.

21 90% say they find TV news reliable; 33% say the same of Internet-based sources: [http://blogs.vancouversun.com/2011/05/11/ubc-study-finds-canadians-trust-mainstream-news-media-more-than-social-networks/]

22 See, for example, Robert W. McChesney, 124. Over the past

twenty years, newsroom staffs have shrunk by 30 to 50 percent across the industry.

23 Wade Rowland, *Greed, Inc. Why Corporations Rule our World and How We Let It Happen*.

24 Ken Auletta 1991.

25 Auletta; Sally Beddel Smith; Edward Bliss; Richard S. Salant; A. M. Sperber.

26 Andrew Hilton, "The Scud Stud Has Come Home."

27 http://www.gallup.com/poll/155585/americans-confidence-television-news-drops-new-low.aspx. This is higher than confidence in the written press at 44%.

28 The polls cannot be directly compared, but the differences in confidence levels are so dramatic that they do support the hypothesis that there is a linkage between the existence/non-existence of competitive public broadcasting and the public's confidence in television news. Polling shows that public trust in television news throughout the EU stands at 61%, a figure that is "a lot lower than it was just ten years ago." (William Horsley, "Public Trust in the Media: Why is it Declining?" Report for the Association of European Journalists.) Canadians are remarkably confident: a 2011 UBC/Angus Reid poll reported that 90 percent of respondents find television news to be "very reliable" or "reliable." [http://blogs.vancouversun.com/2011/05/11/ubc-study-finds-canadians-trust-mainstream-news-media-more-than-social-networks/] (Confidence in newspapers among Canadians was an identical 90 percent.

29 In independent polling, CBC and CBC News Network are consistently ranked first by a wide margin as providing the best national and international news on television, with CTV and CTV News Channel second; CTV tops the rankings for local news. (Anglophone Canadians 18+.) "Who is Canada's News Leader?" *Canadian Media Research Inc.*, March 2012. [http://mediatrends-research.blogspot.ca/2012/03/who-is-canadas-tv-news-leader-who-is.html]

30 Media owners can also be influential in this process, a topic I have dealt with at length in *Greed, Inc.*

31 Frank is quoted in a classic of television scholarship, Edward Jay Epstein, *News from Nowhere: Television and the News,* 40.

32 Clear Channel was purchased in 2008 by Bain Capital, the company founded by 2012 Republican presidential nominee Mitt Romney.

33 Among Anglophones, 18+. Canadian Media Research Inc., 2012

34 http://www.statcan.gc.ca/daily-quotidien/100526/dq100526b-eng.htm

35 Most of this scheduling information has been gleaned from Austin Weir's invaluable and exhaustive first-person account of early public service broadcasting in Canada.

36 Barry Kiefl, Sept. 5, 2012.

37 The four main modes of Internet-facilitated communication are: (1) dialogue — one-to-one communication; (2) information aggregation — the collection of data from many sources by a single distributor; (3) group conversation or fora; and (4) broadcasting — one-to-many communication.

38 In October, 2012, however, the federal government was able to allocate $25 million to the Museum of Civilization to change its name to the Canadian Museum of History and develop an annex displaying Canadian historical artifacts.

39 For an eye-opening look at the blogosphere, see Ryan Holiday's *Trust Me, I'm Lying: the tactics and confessions of a media manipulator* (Toronto, Penguin Group, 2012).

40 Kirstine Stewart, Executive Vice-President, CBC English Services, speech to Canadian Telecom Summit, Toronto, June 6, 2012.

41 Prior to the recent introduction of "portable people meters" in the ratings industry, the average number of viewing hours had been estimated at about 27 hours. The discrepancy results from the fact that people meters measure more than whether a television set is turned on — the measure how many people are watching (or who are at least in the room).

42 "No, Canadians Are NOT Watching Less TV," *Canadian Media Research Inc.*, March 2012. [http://mediatrends-research.blogspot.ca/2012/03/no-canadians-are-not-watching-less-tv.html]

43 "Death of the Canadian Newspaper?" *Canadian Media Research Inc.*, March 2012. [http://mediatrends-research.blogspot.ca/2012/03/death-of-canadian-newspaper]

44 Bell Canada (BCE) purchased CTV, the nation's largest commercial television network, for $1.3 billion in 2010, and formed Bell Media. Bell Media owns 28 television stations and 30 specialty cable channels, including TSN. In 2011 Bell teamed with Rogers Communications to purchase Maple Leafs Sports and Entertainment, owners of Toronto Maple Leafs and other sports properties, for $1.6 billion. Bell also owns a minority stake in the Montreal Canadiens.

45 Interview with Greg O'Brien.

46 Government of Canada. *Broadcasting Act*, 1991, Section 3.

47 A problem associated with leveraged takeovers has been that the enormous debt loads they generate for the purchaser almost inevitably leads to massive layoffs and related service cuts, to reduce costs in the target firm.

48 Standing Committee on Canadian Heritage, *Lincoln Committee Report,* 2003, p. 596. The report commented: "Given that all of these proposed services suit the mandate of a public broadcaster, the Committee cannot understand why the Corporation was denied these services by the CRTC."

49 Concentration of ownership in Canadian commercial media continues apace: in 2012 Bell Media sought to buy Astral Media, operator of 25 television specialty channels including HBO Canada, and 84 radio stations, for $3.4 billion. (The purchase was vetoed by the CRTC and was in abeyance at this writing.) Rogers Media, Canada's biggest cable provider, owns five CityTV television stations in the country's major cities, a clutch of specialty cable channels including Sportsnet, and 51 radio stations. Shaw Media, also a major cable provider, owns the Global television network and 18 cable specialty channels. Quebecor, Quebec's dominant cable provider, in 2000 bought the French language commercial network, TVA and established Quebecor Media, which has since launched the cable channel Sun News. Together these giants claim a substantial majority of the Canadian radio and television

audience.

50 Nordicity, "Value of Public Support for Broadcasters – Simultaneous Substitution and Tax-Based Advertising Incentive," Nov. 2011. Because it airs primarily Canadian programs, CBC television does not benefit to any significant degree from either advertising replacement, simultaneous substitution, or section 19.1 subsidies (and CBC radio not at all).

51 Canadian Radio-television and Telecommunications Commission (CRTC), 2007.

52 Whether commercial and public service media are truly in competition with one another is an interesting question. In one sense they are: if we consider audience numbers to be strictly limited, or "inelastic" as economists put it, then of course people watching public service outlets cannot at the same time be watching commercial offerings. Commercial audience numbers will necessarily drop. But research suggests that audience numbers are not inelastic. There is always a large pool of people who choose to watch or listen to nothing, because nothing on offer interests them. Public service media draw at least some of their audience from that more discriminating pool, and to that extent have little or no impact on commercial media audience numbers. Backing that up, research in Europe has shown that boosting budgets for public service broadcasting has no discernible impact on commercial broadcasting's audiences or revenues. (Irene Costera Mejier, 27-53.)

53 Kiefl, Sept. 5, 2012.

54 Canadian's public broadcaster was initially financed through an annual licence fee for radio receivers that grew to $2.50 by 1953. When television arrived, the anticipated costs were enormous; it was estimated a licence fee of $15 per set was needed to cover them. This was deemed to be politically impossible: Canadians, many of whom were able to receive "free" television signals from American border stations, found it difficult to understand why they should be asked to pay a licence fee. Instead, the radio licence was dropped, and a 15 percent excise tax was imposed on radio and television sets and parts. This brought in more than enough money for the CBC's operations in the first couple of years, but as the market for TV sets became satu-

rated, revenue quickly fell below what was needed even to maintain existing services. (The BBC continues to be financed through a licence fee arrangement.)

55 The media watchdog Friends of Canadian Broadcasting (2004) reports that since 1936, 87 percent of appointees have been affiliated with the governing party.

56 UK *Pilkington Report* (1962), 147, para. 504.

57 In early 2013, CBC management announced it intends to sell *bold*.

58 Kiefl, May 14, 2012.

Works Cited

Ken Auletta, *Three Blind Mice: How the TV Networks Lost Their Way* (New York: Random House, 1991).

Edward Bliss, *Now the News: The Story of Broadcast Journalism* (New York: Columbia University Press, 1991).

Canadian Media Research Inc., "Who is Canada's News Leader?" March 2012. [http://mediatrends-research.blogspot.ca/2012/03/who-is-canadas-tv-news-leader-who-is.html]

_____. 2012. [http://mediatrends-research.blogspot.ca/2012/02/will-cbc-ever-restore-radio-2.html]

_____. "The CBC, ex-CBC Executives and 'Factortion'" Apr. 2012. [http://mediatrends-research.blogspot.ca/2012/04/cbc-ex-cbc-executives-and-factortion.html]

Canadian Radio-television and Telecommunications Commission (CRTC), 2007, "An Overview of the Canadian Program Rights Market," 2b(i).

Irene Costera Meijer, "Impact or Content?" *European Journal of Communications*, 2005, 20.

Senator Pierre De Bané, "The Trouble With Radio-Canada," CTRC Oct. 2012. [https://services.crtc.gc.ca/pub/ListeInterventionList/Documents.aspx?ID=174810&Lang=e]

Deloitte Canada, "The Economic Impact of CBC/Radio-Canada," June, 2011. [http://cbc.radio-canada.ca/_files/cbcrc/documents/latest-studies/deloitte-economic-impact-en.pdf]

Edward Jay Epstein, *News from Nowhere: Television and the News* (New York, Vintage, 1974).

Mark Fowler and David Brenner, "A Marketplace Approach to Broadcast

Regulation." *Texas Law Review*, 60, 207-57.

Todd Gitlin, *Inside Prime Time*, revised edition (New York: Routledge, 1994).

Government of Canada, *Report of the Massey Commission*, "Royal Commission on National Development in the Arts, Letters, and Sciences," 1951. [http://www.collectionscanada.gc.ca/2/5/h5-400-e.html]

Government of Canada. Standing Committee on Canadian Heritage, *Lincoln Committee Report*, 2003. [http://www.parl.gc.ca/HousePublications/Publication.aspx?DocId=1032284&Language=E&Mode=1&Parl=37&Ses=2]

Edward Herman and Noam Chomsky, *Manufacturing Consent: The Political Economy of the Mass Media* (New York, Pantheon Books, 2002).

Andrew Hilton, "The Scud Stud Has Come Home," *Ryerson Review of Journalism*, Spring, 1994.

William Horsley, "Public Trust in the Media: Why is it Declining?" Report for the Association of European Journalists (Linz, November 2008).

Sakae Ishikawa, ed., *Quality Assessment of Television* (Luton: Luton University Press, 1992).

Barry Kiefl, "CBC Cuts: Does Mother Know How to Budget?" Part 1, *Canadian Media Research Inc.*, May 14, 2012.

_____, "Domino Effect Snowballing into a Chain Reaction," *Canadian Media Research Inc.*, Sept. 5, 2012.

_____, "Mr. Bettman, CBC is Losing Money on the NHL," *Canadian Media Research Inc.*, Sept. 15, 2012.

Leslie-Anne Keown, "Keeping up with the times: Canadians and their news media diet," Statistics Canada, 2004. [http://www.statcan.gc.ca/pub/11-008-x/2006008/9610-eng.htm#table2]

Robert W. McChesney, *The Political Economy of Media: Enduring Issues, Emerging Dilemmas* (New York: Monthly Review Press, 2008).

Peter H. Miller, "Developments in the Canadian Program Rights Market," The Canadian Radio-television and Telecommunications Commission (CRTC), 2011.

Geoff Mulgan, *The Question of Quality* (London: The British Film Institute, 1990).

Graham Murdoch, "Public Broadcasting and Democratic Culture," in Janet Wasco, ed., *A Companion to Television* (New York: Wiley, 2009).

Richard Nielsen, "Broadcasting Policy Position Paper" (unpublished) 2012.

Nordicity Research Group, "Analysis of Government Support for Public Broadcasting and Other Culture in Canada." April, 2011. [http://www.nordicity.com/reports.html]

_____. "Value of Public Support for Broadcasters – Simultaneous Substitution and Tax-Based Advertising Incentive," Nov. 2011.

Greg O'Brien. Interview with Robert Rabinovitch. [http://www.cartt.ca/inDepth/details.cfm?featuredNewsNo=1003]

John Reith, *Broadcast Over Britain* (1924) quoted in Asa Briggs and Peter Burke, *A Social History of the Media: From Gutenberg to the Internet* (Cambridge: Polity Press, 2002).

Wade Rowland, *Spirit of the Web: The Age of Information from Telegraph to Internet,* third edition (Toronto, Thomas Allen, 2006).

_____. *Greed, Inc. "Why Corporations Rule Our World and How We Let It Happen,"* third edition (New York, Arcade Publishing, 2012).

Richard Rudin, *Broadcasting in the 21st Century* (New York, Palgrave Macmillan, 2011).

Richard S. Salant, eds. Susan and Bill Buzenberg, *Salant, CBS and the Battle for the Soul of Broadcast Journalism* (Boulder, Westview Press, 1999).

Philip Savage, "Identity Housekeeping in Canadian Public Service Media," in Petros Iosifidis, ed., *Reinventing Public Service Communication* (London: Palgrave Macmillan, 2010).

Roger Silverstone, *Why Study the Media?* (New York: Sage, 1999).

Sally Beddel Smith, *In All His Glory: The Life of William S. Paley* (New York: Simon and Schuster, 1990).

A. M. Sperber, *Murrow, His Life and Times* (New York: Bantam, 1987).

Richard Stursberg, *Tower of Babble* (Toronto: Douglas and Macintyre, 2012).

Michael Tracey, *The Decline and Fall of Public Service Broadcasting* (New York: Oxford University Press, 1998).

Pilkington Report: "Report on the Committee on Broadcasting," Cmnd, 1753 (London: HMSO, 1962)

UNESCO, "Public Broadcasting: Why? How?" 2001. [http://www.unesco.org/new/en/communication-and-information/resources/publications-and-communication-materials/publications/full-list/public-broadcasting-why-how/]

E. Austin Weir, *The Struggle for National Broadcasting in Canada* (Toronto: McClelland and Stewart, 1965).

Acknowledgements

I am indebted to friends and colleagues who were kind enough to read drafts of this manuscript, and whose advice has steered me clear of errors of fact and interpretation: David Skinner, Kealy Wilkinson, Jeffrey Dvorkin, and Barry Kiefl. My brother, Douglas Rowland, as always, provided constructive editing input and encouragement, as did my wife and chief source of inspiration, Christine Collie Rowland. And I owe a debt of thanks to my publisher and editor, Linda Leith, who has had the courage to swim upstream in the current trade book market and publish essay-length works like this one that deal with important public policy issues. A sabbatical from my teaching duties at York University in Toronto has given me the time to research and write this book.

Ot[illegible]s by Wade Rowland

Greed, Inc.: Why Corporations Rule our World

Galileo's Mistake: A New Look at the Epic Confrontation Between Galileo and the Church

Spirit of the Web: The Age of Information from Telegraph to Internet

Ockham's Razor: A Season in France in Search of Meaning

Making Connections: How a Television Team Exposed Organized Crime in Canada

The Plot to Save the World: Stockholm Conference on the Human Environment